DEVELOPING 21ST CENTURY COMPETENCIES IN THE MATHEMATICS CLASSROOM

Yearbook 2016

Association of Mathematics Educators

DEVELOPING 21ST CENTURY COMPETENCIES IN THE MATHEMATICS CLASSROOM

Yearbook 2016

Association of Mathematics Educators

editors

Pee Choon Toh

Berinderjeet Kaur

National Institute of Education

Nanyang Technological University, Singapore

Published by

World Scientific Publishing Co. Pte. Ltd.
5 Toh Tuck Link, Singapore 596224
USA office: 27 Warren Street, Suite 401-402, Hackensack, NJ 07601
UK office: 57 Shelton Street, Covent Garden, London WC2H 9HE

British Library Cataloguing-in-Publication Data
A catalogue record for this book is available from the British Library.

DEVELOPING 21ST CENTURY COMPETENCIES IN THE MATHEMATICS CLASSROOM
Yearbook 2016, Association of Mathematics Educators

ISBN 978-981-3143-60-9

Printed in Singapore

Contents

Chapter 1

21st Century Competencies in Mathematics Classrooms

Pee Choon TOH   Berinderjeet KAUR

This chapter introduces the Singapore Framework for *21st Century Competencies and Student Outcomes* (MOE, 2010) and also provides an overview of the chapters in the book. The chapters are classified into three broad themes. The first is an examination of 21st century competencies and how they can be developed within the context of the mathematics curriculum. The second is an in-depth discussion of evidence-based practices aimed at fostering specific competencies like metacognition and reflective thinking, critical thinking, reasoning and communication skills. The third and last theme is about teaching approaches that are likely to feature increasingly in 21st century classrooms.

1 Introduction

This yearbook of the Association of Mathematics Educators (AME) in Singapore focuses on developing 21st Century Competencies in Mathematics Classrooms. Like previous yearbooks, Mmatical Problem Solving (Kaur, Yeap, & Kapur, 2009), Maathethematical Applications and Modelling (Kaur & Dindyal, 2010), Assessment in the Mathematics Classroom (Kaur & Wong, 2011), Reasoning, Communication and Connections in Mathematics (Kaur & Toh, 2012), Nurturing Reflective Leaners in Mathematics (Kaur, 2013), and Learning Experiences to Promote Mathematics Learning (Toh, Toh, & Kaur, 2014), the theme of this book is shaped by the school mathematics curriculum developed by the Ministry of Education (MOE) and the needs of mathematics teachers in Singapore schools.

The 21st century is often characterised as one where the world we live in is complex, highly interconnected and rapidly changing. The Organisation for Economic Co-operation and Development (OECD, 2005) states that:

> Globalisation and modernisation are creating an increasingly diverse and interconnected world. To make sense of and function well in this world, individuals need for example to master changing technologies and to make sense of large amounts of available information. They also face collective challenges as societies – such as balancing economic growth with environmental sustainability, and prosperity with social equity. (p. 4)

The implication is that students should acquire a set of competencies that would help them better deal with the challenging demands of 21st century life. These competencies go beyond accumulated factual knowledge and relates to how one mobilises cognitive and practical skills, creative abilities, as well as resources such as attitudes, motivation and values, to deal with complex tasks.

In the United States, the vision of the P21 Partnership for 21st Century Learning (Partnership for 21st Century Learning, 2016) since its formation in 2002 is for students to succeed "in a world where change is constant and learning never stops." In their Framework for 21st Century Learning, the mastery of fundamental academic subjects such as Mathematics is identified to be essential for student success. Content mastery should also include the understanding of 21st century interdisciplinary themes like global awareness, financial literacy, civic literacy, health literacy and environment literacy. Building on this bedrock of content knowledge are three types of skills: 1) Learning and Innovation Skills; 2) Information, Media and Technology Skills; 3) Life and Career Skills. In particular, the framework states that:

Learning and innovations skills are what separate students who are prepared for increasingly complex life and work environments in today's world and those who are not. They include:

- Creativity and Innovation;
- Critical Thinking and Problem Solving;
- Communication;
- Collaboration. (p. 2)

In Singapore, the Ministry of Education also introduced its own Framework for 21st Century Competencies and Student Outcomes (henceforth referred to as the Singapore framework) in order to help students "thrive in a fast-changing and highly connected world." (MOE, 2010)

2 Framework for 21st Century Competencies and Student Outcomes

The Singapore framework, shown in Figure 1, is grounded on the belief that knowledge and skills must be underpinned by values. The core values, namely respect, responsibility, integrity, care, resilience and harmony define a person's character and shape the beliefs, attitudes and actions of the person. Thus, these values form the core of the framework. The middle ring represents the Social and Emotional Competencies. These competencies concern firstly with how a student understands and manages him or herself, and subsequently how a student relates to others. The outer ring of the framework represents Emerging 21st Century Competencies that are necessary for success in the globalised world. The framework aims to develop a young person into:

- a confident person who has a strong sense of right and wrong, is adaptable and resilient, knows himself, is discerning in judgment, thinks independently and critically, and communicates effectively.
- a self-directed learner who questions, reflects, perseveres and takes responsibility for his own learning.

- an active contributor who is able to work effectively in teams, is innovative, exercises initiative, takes calculated risks and strives for excellence.
- a concerned citizen who is rooted to Singapore, has a strong sense of civic responsibility, is informed about Singapore and the world, and takes an active part in bettering the lives of others around him. (MOE, 2010)

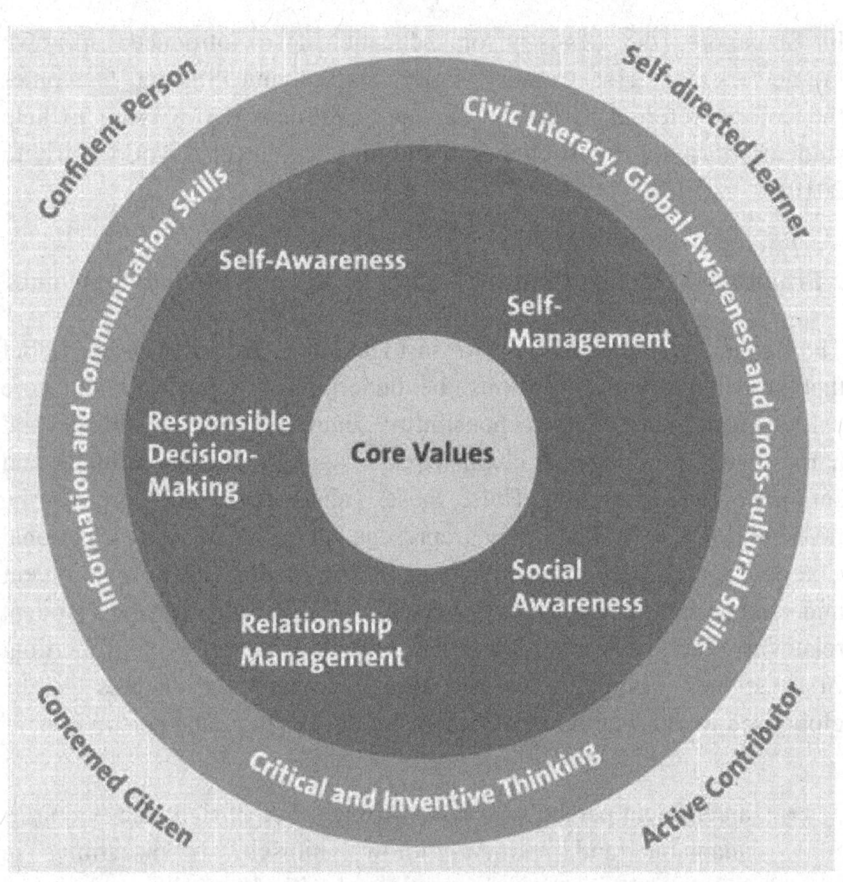

Figure 1. Framework for 21st Century Competencies and Student Outcomes (MOE, 2010)

It is not surprising that the Singapore framework shares many similarities with the Framework for 21st Century Learning from the

United States. One distinct difference is the explicit mention of content mastery as the bedrock of the latter framework. This in no way suggests that 21st century competencies in Singapore is to be developed separately from core curriculum subjects, however it does highlight the difficulty many teachers face in aligning their lessons to the framework. How then should teachers develop 21st century competencies in the classroom? To address the challenge faced by Singapore mathematics teachers, the Association of Mathematics Educators and the Singapore Mathematical Society organised the 2015 annual conference for mathematics teachers with the theme *Developing 21st Century Competencies in the Mathematics Classroom*.

The following 13 peer-reviewed chapters resulted from the keynote lectures and workshops from the conference. The chapters in the book are classified into three broad themes. The first is an examination of 21st century competencies and how they can be developed within the context of the mathematics curriculum. The second is an in-depth discussion of evidence-based practices aimed at fostering specific competencies like metacognition and reflective thinking, critical thinking, reasoning and communication skills. The third and last theme is about teaching approaches that are likely to feature increasingly in 21st century classrooms, for example *flipped classroom* or the use of comics and storytelling. Together, these chapters offer mathematics teachers many examples of how 21st century competencies can be fostered in the classroom.

3 The Mathematics Curriculum and 21st Century Competencies

The aim of the Singapore framework is to develop students into confident persons, self-directed learners, active contributors and concerned citizens. It is reasonable to ask how these outcomes relate to the mathematics curriculum. In chapter 2, Thornton argues that student outcomes can only be achieved through rigorous intellectual engagement with core disciplines. Although learning activities in the 21st century look vastly different from those in the past, mainly because of technology, the core of what makes for good teaching and learning remains largely unchanged. He notes the remarkable similarity of the core values of the Singapore framework with the virtues described by Aristotle more than 2000 years ago and showcases the Polymath project: *Bounded gaps between primes*, as an exemplar of how the core values of respect and integrity is demonstrated through the global collaboration of mathematicians in their quest to solve the *twin prime conjecture*. He goes on to illustrate with a number of local student projects in Australia how the emerging 21st century competencies of civic literacy, global awareness and cross-cultural skills can be fostered in the classroom.

In chapter 3, Wong explores the linkages between the Singapore Mathematics curriculum with the four student outcomes of the Singapore framework. Drawing on several specific real world examples, he discusses how a teacher can help students develop from confident students, to self-directed learners, to concerned citizens and finally active contributors. He ends with suggestions of the competencies teachers themselves should be equipped with in order to model processes required to achieve the desired student outcomes.

Kissane provides an alternative viewpoint in chapter 4 that complements those of the previous two chapters. Instead of focusing on student outcomes, he focuses on the adult roles of 1) the productive worker, 2) the careful consumer, 3) the informed citizen, and 4) the balanced person, that a student assumes after he or she leaves school. He discusses several relevant examples of how mathematics is involved in decisions concerning health, insurance and other aspects of everyday life.

4 Metacognition, Critical Thinking and Communications Skills

The Singapore Mathematics Curriculum framework includes metacognition as one of the five components to help students become better problem solvers. Metacognition is defined as "thinking about thinking", and includes "monitoring of one's own thinking" and "self-regulation of learning" (MOE, 2012, p. 17). Thus it is an indispensable component in developing the desired outcome of a self-directed learner. Kaur, Wong and Bhardwaj presents in chapter 5, an analysis of data collected from forty secondary mathematics teachers involved in the

Teaching for Metacognition project. A pre-intervention survey of the teachers' use of mathematics tasks and understanding of metacognition found that teachers tended to use more performative tasks rather than knowledge building tasks in their classrooms. Performative tasks refer to those that entail the use of lower order thinking skills like recall and direct application of knowledge, and hence are unlikely to encourage students to engage in metacognition. The survey also suggested that the teachers' initial understanding of metacognition was not comprehensive, with many of them associating metacognition with higher order thinking and problem solving. The project thus focused on 1) helping the teachers craft knowledge building tasks, which call for higher order thinking skills, and 2) providing the teachers with strategies to engage their students in metacognition. Examples of how to convert performative tasks into knowledge building tasks, as well as a list of ten learning strategies to encourage metacognition are described in the chapter.

Seto continues with the theme of developing metacognition in the classroom in chapter 6. She focuses on the use of questions to provide opportunities for students to articulate their problem-solving processes and illustrates with two classroom excerpts from the primary level.

Hino describes in chapter 7 two teacher education case studies, and discusses methods for improving teachers' abilities to listen and respond to the often under-developed reflective thinking of young students. The case studies illustrate how tasks from the National Assessment in Japan and results of students' performance on these tasks can be used to help teachers anticipate, compare and investigate students' thinking.

Critical and inventive thinking is one of the Emerging 21st Century Competencies identified in the outer ring of the Singapore framework. One method to foster the development of critical and inventive thinking is the use of open-ended tasks. In chapter 8, Yeo provides a detailed literature review of what constitutes an open-ended task and surveys various research studies involving such tasks. He then provides examples of two types of open-ended tasks that can be employed in the primary classroom and concludes with insights into the various issues that teachers need to be mindful of in the implementation of such tasks.

The ability to communicate is another Emerging 21st Century Competency. In chapter 9, Koay focuses on how to encourage productive mathematics talk in the primary classroom. Useful strategies include establishing classroom routines and supportive environments, the use of rich tasks, the use of talk moves and questioning techniques. She stresses that teachers should relinquish their roles as the sole questioner and assessor of learning in the classroom and allow students to share the responsibility of questioning. This would be a positive step towards helping students take ownership of their own learning and become self-directed learners.

The process of justification involves the two competencies that we have previously discussed, namely critical thinking and communication. In chapter 10, Chua discusses the importance of justification and introduces four types of justification tasks. He then reports on a study of how secondary school students and mathematics teachers perform on justification tasks. The related question of how teachers assess students' justification is also considered in his study. Although the study suggests that many students and even some teachers have difficulties in the process of justification, he believes that given time and sufficient practice, students can become more proficient. To this end, he concludes with five strategies for promoting justification.

While these six chapters focus on developing various competencies, what appears in common is the need for teachers to employ rich mathematical tasks in the classroom. The implication is that teachers should take ownership of their own professional development to update and upgrade their content and pedagogical content knowledge. In doing so, the teachers themselves exemplify the idea of self-directed learners.

Ng and Dindyal proposes that understanding what teachers value when they make decisions in selecting examples for instructional purposes is one way to identify appropriate professional development opportunities. In chapter 11, they report on a study on the use of examples in the teaching of mathematics.

5 Pedagogical Approaches in the 21st Century Classroom

One characteristic of the 21st century is the impact of new technology which is changing the way students learn. A recent phenomenon is the concept of the *flipped classroom*, where instead of learning the content *in class*, students now learn the content *out of class* through either video-recordings or other forms of online learning. Classroom time is then spent on activities like quizzes (to ascertain basic mastery of content and identify misconceptions) and group discussions, which fosters collaborative and communication skills. The flipped classroom is also viewed as an avenue to encourage self-directed learning. In chapter 12, Ho and Chan describe an implementation of the flipped classroom for teaching a group of Junior College students and report on the effectiveness based on the findings of a summative survey, supplemented by student interviews.

In chapter 13, Toh, Cheng, Jiang and Lim introduce a web-based mathematics teaching package that leverages on comics and storytelling. They provide an example of a comic strip that describes two friends on a shopping trip who were puzzled by the % symbol that they saw. Students who followed through the comic strip would be presented with opportunities to learn about percentages and fractions. Toh et al. conclude that one advantage of using comics is that it develops the skill of interpreting real-world information presented in graphic form, another of the Emerging 21st Century Competency identified by the Singapore framework.

Not all learning in the 21st Century need to leverage on technology. In the final chapter, Gura describes a course in game theory that can be adapted to the classroom. Game theory is a branch of mathematics which aims to build models in order to draw conclusions for decision-making. It is part of the content of the Singapore Economics curriculum which states that "one of the 21st Century Competencies emphasised in the Singapore Economics curriculum is sound reasoning and decision making under Critical and Inventive Thinking" (University of Cambridge Local Examinations Syndicate, 2015, p. 2).

6 Concluding Thoughts

21st century classrooms call for 21st century teachers. These teachers must be confident teachers, self-directed learners, active contributors and concerned citizens, in order to be effective role models for their students. In particular, as self-directed learners, they should question, reflect and look for different ways to improve their craft of teaching.

The chapters in this yearbook provide readers and specifically classroom teachers with ideas on the why, what and how of developing 21st century competencies in the mathematics classroom. Readers are urged to read the chapters carefully and try some of the ideas in their classrooms and convince themselves that these ideas offer a means to engage students in meaningful mathematical practices meant to develop the desired learning outcomes.

References

Kaur, B. (2013). *Nurturing reflective learners in mathematics*. Singapore: World Scientific.

Kaur, B., & Dindyal, J. (2010). *Mathematical applications and modelling*. Singapore: World Scientific.

Kaur, B., & Toh, T.L. (2012). *Reasoning, communication and connections in mathematics*. Singapore: World Scientific.

Kaur, B., & Wong, K.Y. (2011). *Assessment in the mathematics classroom*. Singapore: World Scientific.

Kaur, B., Yeap, B.H., & Kapur, M. (2009). *Mathematical problem solving*. Singapore: World Scientific.

Ministry of Education, Singapore (2010). *MOE to enhance learning of 21st century competencies and strengthen art, music and physical education*. Retrieved 31 December, 2015 from www.moe.gov.sg

Ministry of Education, Singapore (2012). *O-Level mathematics: Teaching and Learning Syllabus*. Singapore: Author.

Partnership for 21st Century Learning. (2016). *Framework for 21st Century Learning*. Retrieved 1 January, 2016 from http://www.p21.org/storage/documents/docs/P21_framework_0116.pdf

OECD (2005). *The definition and selection of key competencies: Executive summary.* Retrieved 31 December, 2015 from http://www.oecd.org/pisa/35070367.pdf

Toh, P.C., Toh, T.L., & Kaur, B. (2014). *Learning Experiences to Promote Mathematics Learning*. Singapore: World Scientific.

University of Cambridge Local Examination Syndicate (2015). *H3 Economics examination syllabuses* (for candidates in Singapore only). Cambridge: Author.

Chapter 2

Mathematics Education, Virtues and 21st Century Competencies

Stephen THORNTON

While much research in mathematics education focuses on how best to teach a concept or skills to help students understand or solve a problem, this paper argues that ethical and philosophical decisions about what is worth learning and for what reason are equally, if not more important. The *Singapore Framework for 21st Century Competencies and Student Outcomes* (MOE, 2010) captures some of these ethical and philosophical dimensions in its emphasis on the core values that lie at the heart of the competencies and outcomes. I argue that the outcomes of confidence, self-direction, citizenship and active contribution to society can only be achieved through rigorous intellectual engagement with core disciplines, particularly mathematics. I show how mathematics contributes to and embodies intellectual and moral virtues such as truthfulness, open-mindedness and evidence (Sockett, 2012), and discuss how school mathematics can contribute to the development of competencies such as civic literacy, inventive thinking, and communication skills.

1 Introduction

Much of the current educational rhetoric, particularly in the Western world, emphasizes that society is changing rapidly and that new technologies demand that students in the 21st Century leave school with new sets of skills and competencies. Schleicher (2012), for example, reporting on the *International Summit on Preparing Teachers and*

Developing School Leaders for the 21st Century: Lessons from around the World asks: "What are the skills that young people need to be successful in this rapidly changing world?" The text goes on to argue:

> traditionally mathematics is often taught in an abstract mathematical world, using formalism first, removed from authentic contexts, and discouraging to the students that do not see its relevance – for example, students are taught the techniques of arithmetic, then given lots of arithmetic computations to complete; or they are shown how to solve particular types of equations, then given lots of similar equations to solve. In contrast, in the 21st century, students need to have an understanding of the fundamental concepts of mathematics, they need to be able to translate a new situation or problem they face into a form that exposes the relevance of mathematics, make the problem amenable to mathematical treatment, identify and use the relevant mathematical knowledge to solve the problem, and then evaluate the solution in the original problem context. (p. 34)

In this paper, however, I want to argue that, due in large part to the increased access to modern technology, learning in 21st century mathematics may look superficially different to learning in the past, at heart what matters in good teaching and learning is little different to what has always mattered. In the description of mathematics in the 21st century provided above, relevance to the real world, the ability to formulate a problem mathematically, and the ability to evaluate the solution in the original context, are contrasted with what are termed "traditional" approaches to teaching mathematics. Yet I would argue that good mathematics teaching and learning has always emphasized relevance, problem formulation and real world applicability, as well as stressing the spirit and beauty of mathematics. Similarly, creativity and innovation, critical thinking, problem solving and decision making, and learning to learn and metacognition as described in the *Assessment and Teaching of 21st Century Skills* project (Binkley et. al., 2012) have always been important features of good teaching and learning.

Rather than attempting to list new sets of skills or competencies needed in the 21st Century, which by default tend to devalue the skills and competencies that are assumed to be valued in traditional education,

I argue that a potentially more productive approach is to reflect on the values and ethics that have always underpinned good teaching and learning, and to rediscover these in a 21st Century context. In the next section I discuss the *Singapore 21st Century Competencies* (MOE, 2010) and compare the Core Values at the centre of what has come to be called the "Singapore Swiss Roll" with the intellectual virtues described more than 2000 years ago by Aristotle in the *Nicomachean Ethics* (Irwin, 1999), and given contemporary relevance by MacIntyre (2007) and Sockett (2012). In the following section I describe the *Polymath* twin prime project (Gowers, 2009), using it as a contemporary example of how the discipline of mathematics embodies the virtues of truthfulness, open-mindedness and impartiality described by Sockett. I then give three examples from school mathematics that highlight the competencies of civic literacy, inventive thinking, and communication and information skills described in the Singapore framework. I suggest that the virtues of truthfulness, open-mindedness and impartiality are not only ends in themselves, but underpin each of these three competencies.

2 21st Century Competencies and the Intellectual Virtues

The familiar "Singapore Swiss Roll" describing the desired outcomes of education for Singapore is shown in Figure 1. This framework is designed to develop a young person who is:

- a confident person who has a strong sense of right and wrong, is adaptable and resilient, knows himself, is discerning in judgment, thinks independently and critically, and communicates effectively.
- a self-directed learner who questions, reflects, perseveres and takes responsibility for his own learning.
- an active contributor who is able to work effectively in teams, is innovative, exercises initiative, takes calculated risks and strives for excellence.
- a concerned citizen who is rooted to Singapore, has a strong sense of civic responsibility, is informed about Singapore and the

world, and takes an active part in bettering the lives of others around him. (MOE, 2010)

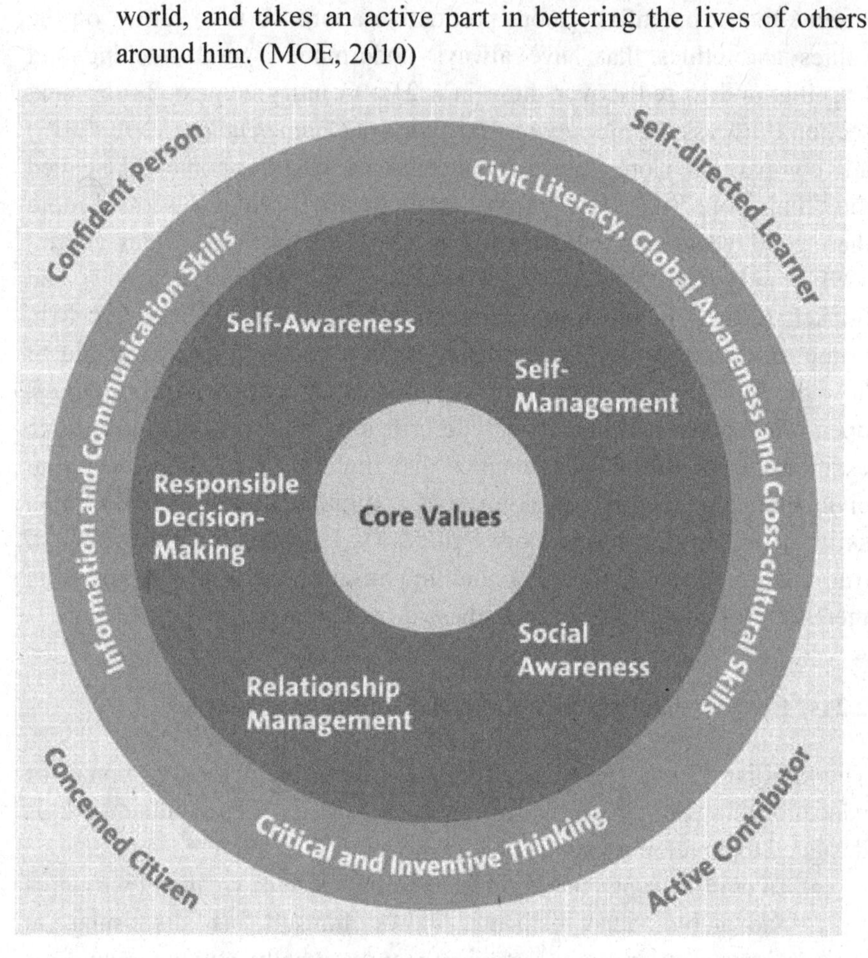

Figure 1. Framework for 21st Century Competencies and Student Outcomes (MOE, 2010)

At the centre of the framework are the core values that describe the person's character, beliefs and attitudes. They are: respect; responsibility; integrity; care; resilience; and harmony, each of which is elaborated upon in more detail. These core values bear a remarkable similarity to the virtues described by Aristotle some 2000 years ago (see Table 1), and are given contemporary credence by writers such as MacIntyre (2007) and Sockett (2012).

In his *Nicomachean Ethics* (Irwin, 1999) dedicated to, or perhaps edited by, his son Nicomacheus, Aristotle articulates those values that are "implicit in the thought, utterance and actions of an educated Athenian" (MacIntyre, 2007, p. 173). Aristotle holds that every activity or enquiry aims at some good, that is what is required to live maximally well, or to attain a state of happiness that he calls *eudaimonia*. However, by happiness Aristotle does not mean the pursuit of material pleasure, but rather a flourishing life, a state of well-being characterized by blessedness and prosperity. It is a state of oneness with self and with society, where the citizen acts virtuously because the act is, in and of itself, virtuous. Aristotle's virtues thus have a universality that transcends place and time, yet the practice of those virtues is characterized by judgment about what is the right thing to do in a particular situation. For Aristotle and indeed Greek society more generally, virtue, or ἀρετή (areté), represents excellence in personal and civic life (Moltow, Thornton, & Kinnear, 2015).

Aristotle distinguishes between the moral virtues, that is those ethical dispositions conducive to living well amongst other persons in society, and the intellectual virtues, that is those intellectual dispositions conducive to the cultivation and deployment of theoretical and practical knowledge. Aristotle holds that the moral virtues that he terms virtues of character are learned through habit and practice, and the intellectual virtues that he terms virtues of thought through instruction. He sees these virtues as inextricably linked; that is, it is not possible to exhibit excellence in personal and civic life without having both moral and intellectual virtues. One classification of Aristotle's virtues is given in Table 1. One can see clearly the core values of respect, embodied in justice and generosity; responsibility, embodied in temperance and prudence; integrity, embodied in truthfulness and wisdom; care, embodied in magnanimity and gentleness; resilience, embodied in courage and harmony; and harmony, embodied in justice and gentleness throughout these virtues.

Table 1
The moral and intellectual virtues

MORAL VIRTUES	INTELLECTUAL VIRTUES
Temperance *(sōphrosunē)*	Intelligence *(nous)*
Courage *(andreia)*	Scientific knowledge *(epistēmē)*
Justice *(dikaiosuné)*	Wisdom *(sophia)*
Truthfulness *(alētheia)*	Art *(technē)*
Generosity *(eleutheriotēs)*	Prudence *(phronēsis)*
Magnanimity *(megalopsuchia)*	
Gentleness *(praotēs)*	

In his seminal *After Virtue*, which has become a key influence in contemporary virtue ethics, Alasdair MacIntyre (2007) argues that the modern project of justifying ethics by appeals to rationality or analysis of linguistic meanings, has necessarily failed. He suggests that contemporary moral philosophy is in a state of crisis, unable to provide a means to settle apparently incommensurable moral dilemmas. His analysis results in a rediscovery of Aristotelian ethics through the articulation of virtues that describe the qualities or attributes that mark a person as one of moral and intellectual excellence.

Hugh Sockett (2012) discusses these virtues in an educational context. He classifies knowledge as public, or propositional, marked by virtues of truthfulness, open-mindedness and evidence, and private, marked by virtues of experience, commitment and identity. While each of these virtues contributes to a person's character, it is the intellectual virtues marking public knowledge that play a particular role in formal education.

Truthfulness, according to Sockett (2012), is a disposition to say what we know is true, and to seek out what is true. It is marked by qualities such as accuracy, care and reliability in acquiring knowledge, and sincerity, the motivation to say what is true. It seems self-evident that mathematics has a particular role to play in the cultivation of truthfulness. Mathematical enquiry is marked by a commitment to accuracy, both in the logical deduction that characterizes activities such as geometric or algebraic proof, and in the evaluation of levels of confidence that characterizes hypothesis testing in statistics. Similarly, mathematics is marked by a commitment to the communication of truth, which is in turn evaluated not by whim or personal preference, but on the basis of the degree to which knowledge claims can be rigorously justified.

Open-mindedness is a disposition to view knowledge as provisional through the consistent construction of alternatives. In his seminal *Proof and Refutations* Imre Lakatos (1976) presents a wonderful example of an open-minded community of mathematical inquiry. The participants in Lakatos's hypothetical classroom consistently question the legitimacy of claims relating to the validity of Euler's Rule linking the number of edges, vertices and faces of a three-dimensional object, in the process refining the definition of a polyhedron and clarifying the conditions under which the rule holds. The participants in the classroom were willing to both suspend judgment and to put aside self-interest in favor of an open-minded commitment to public knowledge. While classrooms such as that described by Lakatos may be hypothetical, there are many similar examples in the literature, such as Brown and Renshaw's (2006) description of collective argumentation in mathematics.

Evidence is characterized by a commitment to justification, impartiality and judgment. Again, I suggest that these characteristics are particularly apparent in the discipline of mathematics. Justification through logical argument is an indispensable aspect of mathematical reasoning, alternative ways of thinking are given intellectual respect, and judgments are made about the truth-value of competing alternatives.

Sockett's (2012) classification of virtues in public knowledge derives from the intellectual virtues described by Aristotle, particularly those of scientific knowledge, wisdom and art. I suggest that there is a remarkable level of congruence between the virtues described by Aristotle and Sockett, and the core values of respect, responsibility, integrity, care, resilience and harmony described in Singapore's model of 21st Century competencies. If we accept that mathematics plays a central role in the development of the intellectual virtues described by Sockett and deriving from those of Aristotle, then it is clear that mathematics is a crucial element in the development of the core values, and hence the realization of the desired outcomes of education, described in the Singapore framework. In the next section I discuss a particular example from the world of mathematics to illustrate in greater detail how it both exemplifies and develops these intellectual virtues.

3 Bounded Gaps Between Primes: The Polymath Project

On 14 May 2013 Yitang Zhang, a then unknown mathematician, published a breakthrough paper (Zhang, 2014) that included a new way of tackling the twin prime conjecture. The conjecture states that there is an infinite number of pairs of primes that are two apart, such as 41 and 43. Zhang announced a proof that there are indeed infinitely many prime pairs, however rather than differing by two as in the twin prime conjecture, his result showed that these infinitely many pairs were no more than 70 million apart. The result was the first time anyone had been able to put a finite bound on gaps between prime numbers, and prompted a flurry of activity around the world. By 30 May 2013 Scott Morrison from the Australian National University announced that he had reduced the gap to 59 470 640, and on 4 June 2013 Terry Tao of the University of California launched a collaborative project as part of the Internet-based *Polymath* endeavour (Gowers, 2009), an open access online collaboration between mathematicians established in order to share knowledge and ideas towards solving previously intractable problems in mathematics. By 27 July 2013 this online collaboration had reduced the gap between prime pairs to 4680. By April 2014 the gap was 246. In less than twelve months more progress was made on the twin prime conjecture than had been made in the previous 2000 years (Nielsen, 2014).

The table of advances published by Nielsen (2014) contains some 300 entries, each of which is the result of work carried out by a mathematician, using variations and ideas built on the initial paper by Zhang. Of these 300 or so entries at least 30 are struck out due to imprecision or recognition of an initial error. Many others are tentative results shown with a question mark.

I suggest that this research provides an example par excellence of the intellectual virtues of truthfulness, open-mindedness and evidence described by Sockett (2012). It shows how the establishment of a cooperative research community enabled the initiation and sustaining of intellectual engagement with a serious problem. It shows how the work of a previously unknown mathematician was recognized by the community, and paved the way for rapid progress. It shows how all results were given intellectual respect, but that each was evaluated according to the standards of rigor established and accepted by the community.

In his blog proposing the *Polymath* project Gowers (2009) provides three arguments for the potential value of massive online collaboration in mathematics. First he argues that if many people are working on a problem there is a greater chance that one person will discover a technique or idea that helps to solve the problem. Second he argues that, because different people know different things, the knowledge of a group is greater than the knowledge of one or two individuals. Third he argues that different people have different approaches to research.

Gowers (2009) goes on to propose some ground rules to guide work on the various problems that are proposed as potential *Polymath* projects. These ground rules include:

- Keep comments short and not too technical. Express ideas in a way that others can build on it.
- Try out lots of ideas, even those not fully thought through.
- Point out errors, but in a respectful way. Do not be self-interested in trying to defend your own ideas.
- Resist the rush towards polished thoughts even if working on the problem individually appears as if it might lead to a solution or justification.
- Only pursue a different approach to a problem or a different problem if the collective agrees that this is a useful thing to do.
- Publication of results is to be in the name of the collective, regardless of the relative proportion of work carried out by individuals.

The *Polymath* project thus provides an excellent example of collaborative problem solving among mathematicians where the intellectual virtues of truthfulness, open-mindedness and evidence are apparent. Although school mathematics can never adopt the same ways of working as academic mathematics, nor be a special kind of academic mathematics (Watson, 2004), it is instructive to ask to what extent school mathematics can contribute to and embed these same intellectual virtues. In the following section I give three examples from school mathematics that illustrate these intellectual virtues and that also help to develop the 21st Century competencies of civic literacy, global awareness and cross-cultural skills; critical and inventive thinking and; communication, collaboration and information skills described in the outer ring of the *Singapore Framework for 21st Century Competencies and Student Outcomes* (MOE, 2010).

4 Examples from School Mathematics

4.1 *Tidal surges: Civic literacy*

The move towards a social perspective on mathematics education, described for example by Lerman (2000) has seen an accompanying trend towards consideration of issues of equity and social justice, described in critical mathematics education (e.g. Frankenstein, 2006; Gutstein, 2003). Critical mathematics education maintains that mathematics has a crucial role to play in engaging with the world of students to highlight issues of physical, economic and social importance (Atweh and Ala'i, 2012). Atweh and Brady (2009) term such an approach socially response-able mathematics, addressing issues of social justice through engagement with mathematics. It is marked by ethical principles that embody respect and concern for the other, and that respond to the demands of the other. Like the intellectual virtues described by Sockett (2012), which precede and underpin our engagement with knowledge, critical mathematics education sees ethical responsibility as a relationship with the other that precedes knowledge of the other.

One particular example described by Atweh and Ala'i (2012) concerns storm and tidal surges in a coastal town in Western Australia. They describe how a teacher who initially felt uncomfortable "imposing" social values on his students was able to make mathematics relevant for students by analyzing a situation of significant importance to the local community.

> The school's mission balances academic excellence and care for inclusion of disadvantaged students, in particular Aboriginal students… [The teacher's] interest in joining the project was due to his concern about making mathematics relevant to students: "to find something that actually meant something to them". Even though he came from a school system that focused on issues of social justice, he did not feel comfortable when dealing directly with issues of social justice in mathematics. In his own words, "if [the activities] go too far [from] the mathematics curriculum - that would worry me". His concern was that such discussion of social justice and values might lead to the imposition of the teacher's values on the students. (p. 101)

However, the teacher planned a school-based project in which students calculated how high the town's storm surge wall needed to be to safeguard the town from the impact of cyclones that frequently hit the area. The major impact of such cyclones is often flooding that occurs at high tide, exacerbated by tidal surges resulting form the cyclone. Students were able to use their knowledge of trigonometric functions and spreadsheets to calculate the combined impact of the two factors. They determined that if a cyclone hit during high tide, the storm surge would be 6.5 meters, whereas the actual height of the existing wall protecting that part of the town is at most 2.7 meters. At the conclusion of the project, the students presented their findings to the local officials and to an engineer who used similar techniques to obtain almost identical results. However, he also pointed out the high cost of building a 6.5 meter wall around the town, the aesthetic impact of blocking people's coastal views, and the problem of water being trapped behind such a high wall. Hence, the local Council had adopted an alternative plan, which was to increase the height from the ground of all new buildings.

What this example illustrates is that concerns for environmental and social issues can be an integral part of school mathematics. Such activities give students a sense of pride and achievement, and also raise their awareness of critical issues in contemporary society. In short, through this activity they become response-able students, developing civic literacy, global awareness and cross-cultural skills. In this example, the intellectual virtues, particularly that of evidence, are also apparent as students sourced critical data, and used that information to analyze a critical problem.

4.2 *Mathematics through art: Critical and inventive thinking*

The potential for developing creativity in school mathematics is, arguably, often overlooked. School textbooks tend to emphasize the skills and techniques of arithmetic and algebra, the properties of geometric figures or the collection and analysis of data. Seldom do they provide students with opportunities to inquire or explore information and ideas, to design and imagine possibilities and actions, to reflect on their thinking and processes, or to synthesize and evaluate a range of possibilities (ACARA, 2013). When creativity is discussed in school mathematics it is often seen as an add-on, perhaps in the context of activities for gifted students, as illustrated by the existence of the International Group for Mathematical Creativity and Giftedness (MCG), affiliated with the International Commission on Mathematics Instruction (ICMI). However, as Sinclair (2009) argues, rather than being frivolous or elitist, considering aesthetic dimensions can be both liberating and connecting experiences in school mathematics.

The following example of the power of mathematics to liberate and connect is taken from *Make it Count*, an Australian Government funded project conducted by the Australian Association of Mathematics Teachers (AAMT) Inc. The aim of the project was to develop an evidence base of practices that improve Indigenous students' learning in mathematics and numeracy through initiatives that were relevant and targeted to their particular situation. One such initiative was that conducted by teachers at Alberton Primary School in South Australia

(Thornton & Statton, 2011). The school is located in a working class area of Adelaide and has a large proportion of students from low socioeconomic backgrounds, including a significant number of students from Indigenous backgrounds. Motivated by a student who asked "Why can't we draw all day?", teachers chose an area of interest and developed numeracy activities focused around that interest. Students then had the freedom to choose which group they joined. Areas of interest included marine science, art, popular culture, or architecture.

Thornton and Statton (2011) report how one teacher, Aaron, was a talented artist who had always expressed a personal dislike for mathematics. Aaron reported that he could not see the connection between mathematics and art, however through exploring ideas such as the golden ratio, he was able to overcome many of his misapprehensions and find meaningful situations through which he could integrate mathematics into art.

> In one lesson he worked with a mixed group of year 3, 4 and 5 students to mathematise the process of drawing a face. The students took measurements in real contexts of the position of various facial features, and calculated key proportions. They also looked at the overall shape of faces, noting the variation from a circle or ellipse. The students then drew an outline of a face, and used rulers to measure and calculate the required positions and size for the eyes, ears and nose for their own outline. (Thornton & Statton, 2011, p. 4)

At the conclusion of the mathematics and art activity, Aaron's students gave an exhibition at which they displayed their art and talked about the mathematics they had learned through art. Figure 2 shows one example of a boy who made 3-dimensional sculptures and was able to talk about the number of edges, faces and vertices, and, eventually about Euler's Rule, the relationship between them.

Figure 2. Artwork developed in an integrated art and mathematics activity

The deliberate embedding of mathematics within art made mathematics come alive for both Aaron and his students. Students saw mathematics as a rich area of learning as they extended their own understanding of measurements, proportions and geometry, and as being useful in authentic situations such as drawing more realistic faces or creating 3-dimensional sculptures. Far from being frivolous or elitist Aaron's approach of emphasizing creativity and innovation became central to the students' learning of mathematics and proved to be particularly beneficial to these students who had previously disliked mathematics or struggled to see its relevance in their lives. By opening the possibility of learning mathematics through art, which in turn gave access to mathematical learning to students previously disengaged or alienated from mathematics, Aaron's approach further embodied the open-mindedness that Sockett (2012) lists as an intellectual virtue.

4.3 *Froebel and Scratch: Information and communication skills*

Creativity and innovation are at the heart of the philosophy of Friedrich Froebel (1782-1852), the founder of the kindergarten movement (Manning, 2005), who became a major influence in the pedagogical approach later developed by Maria Montessori and the architectural designs of Frank Lloyd Wright. In contrast to the rigid, passive and repetitive nature of school prevailing at the time, creativity and experiential learning were at the heart of Froebel's methods. He emphasized children's uniqueness in their learning styles; he emphasized the importance of play not only for learning but as an end in itself; and he emphasized the need for children to construct meaning through the manipulation of physical objects. All this was guided by the teacher through the thoughtful and systematic use of carefully designed learning materials, many of which have a mathematical basis. Froebel described his learning materials as "gifts" and the accompanying activities as "occupations". The gifts included soft colored balls, wooden spheres, a cube that breaks into 27 smaller cubes, a cube, sphere and cylinder hanging from a bar, and blocks of various lengths, perhaps the forerunner of Cuisenaire rods.

Froebel held that learning happens first through general impression, then perception by looking at a single object, then perceiving qualities and relations, then comparing, judging and finally making conclusions (Wiebé, 1869). Such activities are clearly impossible without high levels of verbal and visual communication, both between children themselves and between children and their teacher. His occupations therefore included paper-folding, weaving, construction and drawing. Of course, they did not include use of ICT as Froebel lived long before the advent of modern technology, but one can easily relocate many of his suggested occupations into the 21st Century.

One of these was creating designs using lines, for which Froebel used a slate etched with horizontal and vertical lines. He advocated that children first practice drawing horizontal lines, then vertical, then combinations of both, then diagonal lines, and presented several complex designs that could be constructed in this way. One such design is shown in Figure 3.

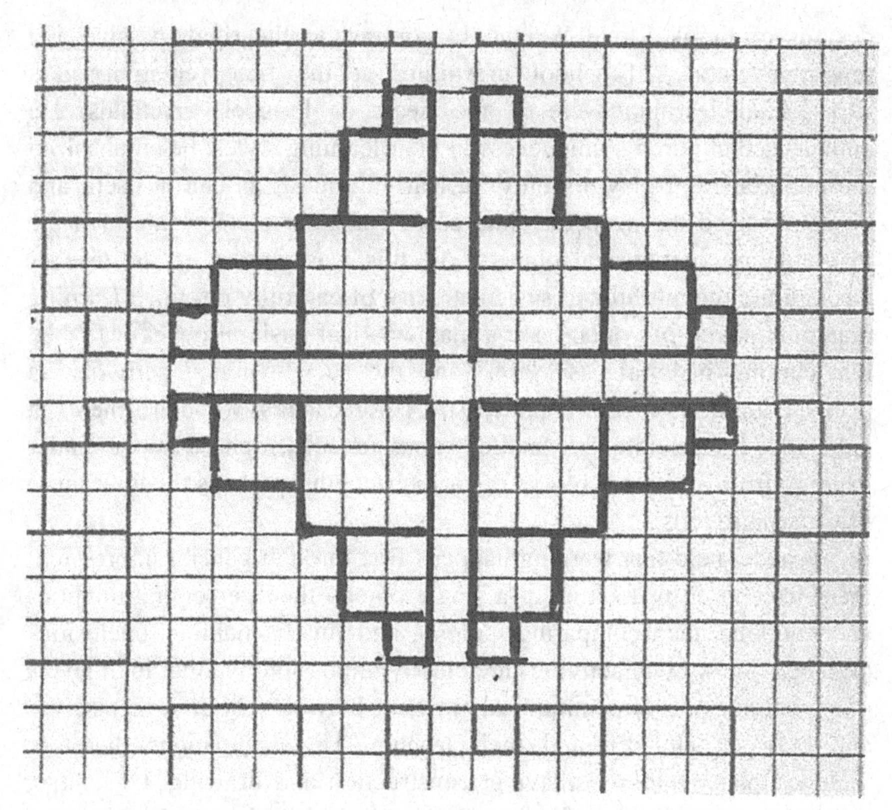

Figure 3. A horizontal and vertical line design suggested by Froebel (Wiebe, p. 196)

Although designed originally for young children to practice drawing, such a design could be used with older children to develop their skills in communication and coding, particularly through the use of Scratch. I created a Scratch program to create this design with the possibility of extension to more than three squares in each quarter and automatic scaling to fit the drawing screen. The coding is shown at https://scratch.mit.edu/projects/64324332/.

Although the design in Figure 3 could be constructed with a combination of turns and moves, generalizing it for any size design requires the careful use of coordinates, the definition and use of algebraic variables to represent lengths of lines and numbers of squares and to enable the calculation of appropriate lengths, as well as careful sequencing and looping to create consistent code. Although using Scratch to create this design might appear to be an exercise that is peripheral to the curriculum, it involves a number of key mathematical concepts learned in secondary school. In addition, it involves high levels of communication and develops coding skills, an integral aspect of ICT literacy.

Activities such as this both highlight the information and communication skills described in the Singapore Framework and develop a concern for accuracy described by Sockett (2012) as part of the intellectual virtue of truthfulness. The act of coding demands attention to precision and offers opportunity to gain instant feedback in a low-cost environment.

5 Conclusion

More than 2000 years ago, in his *Nicomachean Ethics*, Aristotle (Irwin, 1999) described a set of intellectual and moral virtues, which bear a remarkable resemblance to the core values articulated in the *Singapore Framework for 21st Century Competencies and Student Outcomes* (MOE, 2010). In this paper I have argued that mathematics plays a crucial role in the development of the intellectual virtues of truthfulness, open-mindedness and evidence described by Sockett (2012) and built on the ideas of Aristotle. Furthermore school mathematics can play a crucial role in developing the competencies of civic responsibility and global awareness, of critical and inventive thinking, and of communication and information skills. Aristotle also described the joy of mathematics as lying in its inherent order and symmetry. I cannot help but question whether the current emphasis on testing and test results that is so much a part of modern education systems, and that arguably narrows the curriculum to a set of easily measured skills, is removing a lot of that joy and beauty from school mathematics. In the process the emphasis shifts from what it means to be an educated citizen, that is one who displays the intellectual virtues, towards learning only what it takes to be successful at a relatively narrow range of school assessments.

Acknowledgements

I wish to thank my colleagues Dr David Moltow from the University of Tasmania and Dr Virginia Kinnear from Flinders University for their frequent discussions of virtue ethics and mathematics, and for the stimulus to explore this area.

References

Atweh, B., & Ala'i, K. (2012). Socially Response-able Mathematics Education: Lessons from Three Teachers. In J. Dindyal, L. P. Cheng, & S. F. Ng (Eds.), *Mathematics education: Expanding horizons (Proceedings of the 35th annual conference of the Mathematics Education Research Group of Australasia)* (pp. 98-105). Singapore: MERGA.

Atweh, B., & Brady, K. (2009). Socially response-able mathematics education: Implications of an ethical approach. *Eurasia Journal of Mathematics, Science & Technology Education, 5*(3), 267-276.

Australian Curriculum and Assessment Reporting Authority [ACARA]. (2013). *Critical and creative thinking.* Retrieved 29 September, 2015 from http://www.australiancurriculum.edu.au/generalcapabilities/critical-and-creative-thinking/introduction/introduction

Binkley, M., Erstad, O., Herman, J., Raizen, S., Ripley, M., Miller-Ricci, M., & Rumble, M. (2012). Defining twenty-first century skills. In P. Griffin, B. McGaw and E. Care (Eds.) *Assessment and teaching of 21st century skills* (pp. 17-66). Dordrecht: Springer.

Brown, R., & Renshaw, P. (2006). Transforming practice: Using collective argumentation to bring about teacher change in a year 7 mathematics classroom. In P. Grootenboer, R. Zevenbergen, & M. Chinnappan (Eds.) *Identities, cultures, and learning spaces: Proceedings of the 29th annual conference of the Mathematics Education Research Group of Australasia* (pp. 99-106).

Frankenstein, M. (2006). Reading the world with math. In E. Gutstein (Ed.) *Rethinking mathematics: teaching social justice by the numbers.* Wisconsin: Rethinking Schools.

Gowers, T. (2009). *Polymath Project*. Retrieved 29 September, 2015 from http://polymathprojects.org/

Gutstein, E. (2003). Teaching and learning mathematics for social justice in an urban, Latino school. *Journal for Research in Mathematics Education, 34*(1), 37-73.

Irwin, T. (1999) (trans.). *Aristotle: Nicomachean Ethics*. Indianapolis, IN: Hackett.

Lakatos, I. (1976). *Proofs and refutations: The logic of mathematical discovery* (J. Worrall, & E. Zahar, Eds.). Cambridge, UK: Cambridge University.

Lerman, S. (2000). The Social Turn in Mathematics Education Research. In J. Boaler (Ed.), *Multiple Perspectives on Mathematics Teaching and Learning* (pp. 19-44). Westport, CT: Ablex Publishing.

MacIntyre, A. (2007). *After Virtue (3rd ed.)*. Notre Dame: University of Notre Dame Press.

Manning, J. P. (2005). Rediscovering Froebel: A call to re-examine his life & gifts. *Early Childhood Education Journal, 32*(6), 371-376.

Ministry of Education, Singapore (2010). *MOE to enhance learning of 21st century competencies and strengthen art, music and physical education.* Retrieved 29 September, 2015 from www.moe.gov.sg

Moltow, D., Thornton, S., & Kinnear, V. (2015). *Mathematics education as a practice in pursuit of [intellectual] excellence.* Paper presented at the 39th annual conference of the International group for the Psychology of Mathematics Education, Hobart 13-18 July, 2015

Nielsen, M. (2014). *Polymath8 Project*. Retrieved 29 September, 2015 from michaelnielsen.org/polymath1/index.php?title=Bounded_gaps_between_primes

Schleicher, A. (ed.) (2012). *Preparing Teachers and Developing School Leaders for the 21st Century: Lessons from around the World.* Paris: OECD publishing.

Sinclair, N. (2009). Aesthetics as a liberating force in mathematics education? *ZDM: International Journal on Mathematics Education, 41*(1-2), 45-60.

Sockett, H (2012). *Knowledge and virtue in teaching and learning: The primacy of dispositions.* New York: Routledge.

Thornton, S., & Statton. J. (2011). *Mathematising and contextualising - connecting mathematics and numeracy to improve learning for Aboriginal students (Make it Count)*. Paper presented at the 3rd International Realistic Mathematics Education Conference, Boulder. CO: September 23 - 25, 2011

Watson, A. (2004). Affordances, constraints and attunements in mathematical activity. In O. McNamara, & R. Barwell (Eds.), *Research in Mathematics Education, Volume 6: Papers of the British Society for Research into Learning Mathematics* (pp. 23-34). London: British Society for Research into Learning Mathematics.

Wiebé, E. (1869). *The paradise of childhood: a manual for self-instruction in Friedrich Froebel's educational principles, and a practical guide to kinder-gartners.* M. Bradley & Company.

Zhang, Y. (2014). Bounded gaps between primes. *Annals of Mathematics, 179* (3), 1121-1174.

Chapter 3

Enriching Secondary Mathematics Education with 21st Century Competencies

WONG Khoon Yoong

Typical mathematics lessons tend to focus on mastery of specific procedures through drills in routine problems, and this type of learning experience is not motivating for many secondary school students. One way to enrich their learning is to show them how the assigned learning tasks can help them develop 21st century competencies (21CC). In Singapore, 21CC covers four types of student outcomes: confident person, self-directed learner, active contributor, and concerned citizen. These desirable characteristics are strongly related to the Singapore curriculum framework. Confidence can be cultivated through exercises in critical thinking; self-directed learning through taking responsibility of one's learning and engaging in metacognitive reflection; concerns as Singapore citizens through working on mathematics applications and modelling with Singapore contexts; active contribution is made through team work to achieve common goals. This paper elaborates on these outcomes with several mathematics examples.

1 Introduction: Necessity for 21CC

Under traditional direct instruction, teachers spend substantial amount of class time on showing how to carry out standard procedures, for example, to multiply decimals, factorise algebraic expressions, and solve trigonometric equations. The students will use the remaining class time to practise these skills on routine problems. This drill-and-practice approach helps students to master standard skills in accuracy and speed, but many secondary school students find this type of learning boring and demotivating. One way to enrich this routine learning experience is to let students work on tasks that can help them develop 21st century competencies (21CC).

We are now one and a half decades into the 21st century, and the first generation of 21st century students, who are 12+, are already in secondary schools. Yet 21CC hardly form the core of many school subjects, including Mathematics. This raises the critical question of whether the current school experiences are adequate in preparing these students for future learning and career. That future has been described as *globalised, digital, knowledge-based, uncertain,* and so forth. A global trend is to include 21CC into the school curriculum. For example, the *P21 Partnership for 21st Century Learning* (Partnership for 21st Century Learning, n.d.) from the United States aims to ensure that "all learners acquire the knowledge and skills they need to thrive in a world where change is constant and learning never stops". The Singapore Ministry of Education also recognises the need to prepare students to seize the opportunities brought about by globalisation and technological advances (MOE, 2010). The continuing prosperity of the nation depends on whether or not students acquire 21CC now as well as into the future through life-long learning.

In Singapore, the typical approach to inculcate 21CC is through active participation in co-curricular activities (CCA), community services projects, and sporting events. The challenge faced by mathematics teachers is how to complement these traditional activities with mathematics learning experiences that can actively foster 21CC.

2 21CC and the Singapore Mathematics Curriculum

21CC is a multi-dimensional construct, and it is often related more to so-called soft skills or life skills than to technical or academic competencies. Its main components are given in Table 1.

Table 1
Common components of 21CC

Components	Remarks
Technology	Digital literacy, social media
Cognition	Critical thinking, creativity, financial literacy, multi-disciplinary processes
Metacognition	Learn to learn
Emotions	Grit, resilience, growth mindset, emotional intelligence, risk taking
Values	Fairness, kindness, honesty
Social relationships	Communication, collaboration, social intelligence, online networking, global awareness
Health literacy	Knowledge of health information, personal responsibility

In fact, many of these competencies are not new. They have been expounded in previous centuries by educators and laypeople. For example, the late prime minister of Singapore, Mr Lee Kuan Yew, in a speech given in 1967, noted that

> We want our schools to produce citizens who are healthy and hardy, with a sense of social purpose and group discipline, prepared to work and to pay for what they want, never expecting something for nothing. … [Our schools] will teach our students high standards of personal behaviour, social norms of good and bad, right and wrong. (Lee, 2013, p. 151)

In a recent OECD report (Hanushek & Woessmann, 2015), Singapore education was ranked top of 76 education systems worldwide. However, in a local press report (Teng, 2015), the coordinator of PISA, Mr Andreas Schleicher, was quoted as commenting that "Singapore may need to put greater emphasis on students developing creativity, critical thinking and collaborative skills, and build character attributes such as mindfulness, curiosity, courage and resilience". Indeed, these are important 21CC to foster among students at all levels.

The recent framework for 21CC by the Singapore Ministry of Education (MOE, 2010) identifies detailed attributes leading to four types of student outcomes: confident person, self-directed learner, concerned citizen, and active contributor. These outcomes can be linked to the various components of the Singapore Mathematics Curriculum Framework, as shown in Table 2.

Table 2
Correspondence between 21CC student outcomes and the Singapore mathematics curriculum

21CC: Student Outcomes	Components of Singapore Maths Curriculum
Confident person	Attitudes, Concepts, Skills, Processes
Self-directed learner	Metacognition, Self-regulated learning
Concerned citizen	Attitudes, Processes
Active contributor	Processes, Problem Solving

This correspondence between the two frameworks suggests that implementing 21CC does not require making drastic changes to mathematics lessons. What might be needed is to enrich current learning experiences to cover these four student outcomes. Each of these outcomes will be discussed separately in the next four sections, following a progression from *confidence* to *self-directedness* to *concerns* and finally to *activeness,* in ever deepening cycles with topics and grade levels. In the penultimate section, it is suggested that these four outcomes also apply to mathematics *teachers* of the 21st century.

3 Confident Person

Confident students are those who are able and willing to apply whatever they have learned to challenge ideas that appear to be wrong or counter-intuitive rather than to accept them as given "facts". This critical thinking is a powerful attribute of 21CC because a confident person in the 21st century must be able to evaluate the validity and implications of numerous ideas and "facts" he or she will encounter in the globalised, digital world. Mathematics learning is particularly effective in dealing with numbers, logic, and diagrams, which people tend to indiscriminately accept as true. To counter this unquestioned tendency, critical thinkers will first pose their own questions about the given information. Next, they will try to answer the questions by conducting research, collecting relevant information, and discussing their ideas with other people. Quite often, there may not be right-or-wrong answers, but the *process* of trying to find answers rather than being *told* "the" answers provides the necessary training to develop 21CC. This is illustrated below by an example about risks.

The mass media contain numerous reports about health risks. "Risk" is not a topic in the current mathematics curriculum, although basically it means the probability of suffering from loss or harm. Consider this finding. A report (Pang, 2012) noted that among Chinese-Singaporeans, those who ate Western fast food more than four times a week had an 80% increase in the risk of dying from heart disease. Of 811 people who followed this diet, 17 died from heart disease. Given this information, the student can compute the original risk ($^{17}/_{811}$ or about 0.02), and work out how many more people would die due to the 80% increase in risk. On the basis of this analysis, one can make an informed judgment about whether the "substantial" percentage increase as reported is misleading or really "serious". Continuing along similar probing will lead to questions about risks due to differences in age, gender, socio-economic status, and other factors. Eventually this may lead to changes in one's eating habits. For instance, the report mentioned that Health Promotion Board recommends that people should not eat fast food more than twice a week (Pang, 2012). This brief outline shows how learning activities can be planned to promote critical thinking.

The confidence in independent thinking and search for answers that require mathematics does not develop readily among students who practise solving only routine problems, because routine practice tends to cultivate a compliant attribute: they are good at carrying out instructions but may not be able to think for themselves, without first seeking "permission" or guidance from others such as teachers and those in authority. This inability to think for oneself under unpredictable situations is sadly illustrated by the Sewol tragedy that happened in

Seoul on 16 April 2014. When the Sewol ferry was sinking, most of the student passengers obeyed the announcement to stay in the cabin (Sinking of MV Sewol, n.d.), instead of using whatever information was available to make their own decision for survival in a turbulent situation.

The training for independence and confidence in one's own thinking can begin in the safe classroom environment. The first type of learning task is to ask students to pose and solve their own problems. In his doctoral study, Chua (2011) found that Secondary 3 students were generally not competent in posing meaningful geometry problems: about 20% of their problems were vague or mathematically wrong and nearly half the solvable problems involved only direct recall. This weakness in problem posing and asking questions in general needs to be addressed (e.g., Brown & Walter, 2005; Wong, 2015b). This "questioning competency" can be applied to tackle open-ended investigations and to make decisions about real-life issues based on available information and mathematical thinking and modelling (e.g., Lee & Ng, 2015). A modelling example involving investment is given below. It can be used to enhance financial literacy, which is another important component of 21CC.

Assume that Bill has bought 1000 shares of a company at $1 each. As the share market tumbled, he bought and sold the shares in several transactions, and finally managed to sell all the shares when it went back to $1 each. Did he break even? Why or why not? To answer this question, students have to make some assumptions, pose questions about specific transactions, compute changes in Bill's investment, look for patterns, and evaluate the effects of his strategy. They will engage in simulations and thinking about plausible outcomes. Table 3 shows one way to keep track of four possible transactions. Instead of asking students to complete this table, let them design their own way to monitor the transactions, because this is an important step they ought to learn when they solve real-life problems. The transactions given in Table 3 show that Bill has made a net profit of $50. Of course, other outcomes are possible due to the unpredictable nature of the stock market, and different approaches proposed by the students should be discussed.

Table 3

Transactions for Bill's investment

Transactions	Cash ($)	Number of Shares Held
Start	1000	0
Buy at $1	0	1000
Sell at 90¢	900	0
Buy at 85¢	50 (900 – 850)	1000
Sell at $1	1050 (50 + 1000)	0

4 Self-Directed Learner

A self-directed learner can take ownership of one's learning by developing a set of learn to learn skills and habits. Self-directed learning is a key component of metacognition, one of the five factors in the Singapore Mathematics Curriculum. The National Research Council (2005) in the United States referred to this aspect of metacognition as helping "students learn to take control of their own learning by defining learning goals and monitoring their progress in achieving them" (p. 2). Several components of metacognition are discussed in the literature (e.g., Buoncristiani & Buoncristiani, 2012; Efklides, 2006; Flavell, 1976; Schraw & Moshman, 1995), and four of these components are selected here as especially relevant for promoting self-directed learning (see Wong, 2015a):

- Metacognitive knowledge (MK): knowledge about effective and ineffective learning strategies.
- Metacognitive skills (MS): actual competence in carrying out specific learning strategies.
- Metacognitive monitoring (MM): be mindful of what happens when certain learning strategies are put into practice; aware of one's perceptions of the learning tasks as easy or difficult, familiar or novel, and so on.
- Metacognitive reflection (MR): think about the effects of metacognitive practices and what to change to bring about better learning in the future.

Without proper guidance, students may not be able to develop effective learning strategies on their own. Mathematics teacher education programmes seldom cover these strategies, and research about study methods for mathematics is quite limited, although advice about these methods has been offered (e.g., Ooten, 2010). From personal encounters, many teachers are not able to advise their students about alternative learning strategies, other than more routine practice. Indeed, "practice makes perfect" seems to be the predominant belief about how to improve performance (e.g., Kaur, 2011; Wong & Tiong, 2006). However, when students succeed in solving simple problems and have filed their worksheets neatly, they may fall into what Karpicke, Butler and Roediger (2009) called the *illusions of competence* (also see Oakley, 2014). These observations support the assertion made by Buoncristiani and Buoncristiani (2012) that "students make poor choices about how they study because they are unaware of alternatives and their consequences" (p. 128). On a positive note, Wiliam (2011) claimed that students can be helped to improve their learning. Hattie (2009) urged teachers to understand "the critical role of teaching appropriate learning strategies" (p. 36). Hence, there is a need for teachers to help students develop more effective learning strategies.

Wong (2013) discussed four ways to help students improve their learning: homework log, records of mistakes, concept map, and student question cards. What follows below are two different strategies that cover the four types of metacognitive attributes mentioned above: *distributed practice* and *learning outcomes*.

Example 1. Distributed practice.

For metacognitive knowledge (MK), explain to students that there are different types of practices and they are not equally effective. If the students *practise* skills wrongly without immediate feedback, they will be *perfect* in reproducing those mistakes, and this is definitely not the desirable outcome. Research (e.g., Rohrer, Dedrick, & Burgess, 2014) has shown that practising a specific number of problems in one setting (*block practice*) is not as effective as practising a few of them at initial learning and then working on the remaining problems at different practice sessions over weeks or months (*distributed* practice), with adequate feedback in both cases. To develop the related metacognitive skills (MS), the teacher will divide a set of practice problems into a schedule according to the stipulated guideline and guide students through it. For metacognitive monitoring (MM), students will record how closely they have followed the schedule and their own progress through the problems. Finally, they reflect on their experiences with distributed practice (MR). This process should be repeated for different topics using different practice schedules to enable students to compare and contrast their experiences under a variety of learning situations. In the end, students learn to regulate their own practice.

Example 2. Learning objectives or specific instructional objectives.
After graduating from pre-service training, many teachers I have worked with ignore the important roles of learning objectives:

- Most of them do not write down these objectives.
- Those who are expected to do so by their school write single, broad objectives, for example, to solve linear equations, without breaking this skill down into a hierarchy of sub-skills which are to be taught systematically in different episodes of the lessons.
- Most teachers do not state learning objectives at the beginning of their lessons. A few teachers read the objectives in passing or flash them on PowerPoint slide so quickly that most students ignore this teaching episode.
- No teacher has been observed to use learning objectives as a tool to monitor the progress of the lesson.

However, self-directed learners must be able to define realistic learning goals and monitor how well they have achieved them. This competency requires training, and one strategy is to show students how to use learning objectives to guide their learning. A sketch of how this might be done is described below.

First, the teacher needs to realise that informing students about learning objectives can motivate them to pay attention. This idea is explained to the students as metacognitive knowledge (MK). Write on the whiteboard the learning objectives, stated in student-friendly words, and spend a few minutes elaborating on them or encouraging students to share their views about the objectives. It might be helpful to get students to copy these objectives into their notebook or worksheets. This is part of metacognitive skill (MS). As the lesson progresses, the teacher will tick off the objectives when they have been dealt with, and this helps the students to engage in metacognitive monitoring (MM). At the end of the lesson, the teacher can ask students which objectives they have understood and which ones need further elaboration. This can be done verbally or using exit cards. This allows students to engage in metacognitive reflection (MR). As before, this training has to be repeated for many lessons before its benefits can be realised.

5 Concerned Citizen

According to the Singapore 21CC framework (MOE, 2010), a concerned citizen

> is rooted to Singapore, has a strong sense of civic responsibility, is informed about Singapore and the world, and takes an active part in bettering the lives of others around him [*sic*].

This includes an awareness of events in both Singapore and the world in order to inculcate global awareness. Mathematics problems can include community, national, and global contexts, but more importantly, solving these tasks must help students gain new knowledge about the world around them, which should be a critical curriculum goal for the 21st century (Wong, 2015a). This aligns with an assumption made by Mason, Burton, and Stacey (2010) that "mathematical thinking helps in understanding yourself and the world" (p. x).

Consider the following problem with a "water conservation" context:

> Mr Tan intends to reduce his current daily water usage by 5%. His current daily usage is 176 litres. How much water will be saved if he is successful in his plan?

In general, students just solve it as a standard percentage problem and do not think much about the answer. They do not learn about the needs to conserve water, the economic benefits of doing so, the water sources in Singapore, the blight of water shortage, and global concerns about water shortage. Without gaining new knowledge, the students are unlikely to develop any concern about how these issues might affect their life and that of others now and into the future. Mathematics teachers should not be fully satisfied with just getting correct answers from their students to routine "real-life" problems; they should encourage their students to think about the deeper significance of these real-life contexts. They need to work with teachers of other subjects to design multi-disciplinary tasks to provide the opportunities for students to learn to integrate knowledge and skills from different disciplines to tackle significant issues.

Many contemporary issues can be weaved into problem contexts, such as healthy diets, energy conservation, recycling, kindness, and so forth. Examples can be found in Lee and Ng (2015), Wong (2003, 2015a), Wong et al. (2012), and online resources, e.g., the *RadicalMath* website (http://radicalmath.org). The following section shows how a mathematical lens can illuminate different notions about *fairness*.

Fairness is an important social concept, which impacts on the behaviours and beliefs of people of all ages and from all walks of life. To embed this concept into mathematical problems, let students discuss their views about fairness in each of the following situations, when a sum of $30 is to be divided among two persons *A* and *B*:

1. Each person gets an equal share, i.e., $15 each.
2. The money is earned from a joint venture, and the money is to be divided according to individual's effort put into the venture, e.g., the effort is 70% versus 30%.
3. The money is given out as vouchers to help people who need it, e.g., poorer people will get a larger share, say *A* earns half as much as *B*.
4. Fairness as "envy free", that is, people get what they want. Let *A* split the amount in a way that he thinks is fair and then let *B* choose first. For money, this usually ends up with equal sharing.

However, different answers may be obtained if it is about sharing a cake!

5. Play the "ultimatum" or money-splitting game (Ultimatum game, n.d.). *A* and *B* are strangers and do not communicate during the game. Let *A* propose how to divide the sum between him and *B*. *B* decides whether to accept or reject *A*'s proposal. If *B* accepts the proposal, the money is split accordingly; otherwise, neither of them receives any money. This game is to be played only once. How do the students react to different splits, say 50-50, 70-30, and so on? Do their reactions change if *B* knows that *A* is a computer?
6. How do the above strategies work if there were three or more students and the amount is larger?
7. Consider fairness in a real-life situation. The price of a face mask is $10. During prolonged hazy days, the shop raises the price to $15. Is it fair? The students may conduct a survey and compare the findings with their own values and ideas about fairness.

According to Brafman and Brafman (2008), there are cultural differences in interpreting fairness, and knowing about these differences in different social concepts becomes increasingly important in a globalised world. In Singapore, students routinely work with classmates from different ethnic and religious backgrounds, and the annual commemoration of the International Friendship Day has exposed them to many different cultures. Contents from ethno-mathematics and multi-cultural traditions can further strengthen this cultural understanding (e.g., Nelson, Joseph, & Williams, 1993). However, as students from different backgrounds may hold strong views about these social-cultural-economic issues, teachers who intend to use these tasks must be able to help students navigate through these differences based on rational thinking rather than emotion-laden responses.

6 Active Contributor

Students who have developed concerns about specific issues should be encouraged to take active steps to address those concerns in order to improve the situations. In the words of creativity author Berger (2014), asking questions "without taking any action may be a source of stimulating thought or conversation, but it is not likely to produce change" (p. 31). However, like other types of learning, to take actions requires motivation, practice, feedback, and encouragement. A student can be an active contributor individually or in a team.

As an example of being an active contributor to personal needs, consider the water saving problem above. Students may measure their own water usage, collect information about different water saving devices and practices, implement small-scale changes, and determine the effects of their actions. For example, if they learn that reducing shower time by one minute can save about six litres of water, they may check this out through own experimentation and contribute through individual actions to national policies.

Students can learn to be active contributors when they work in teams on extended mathematics investigations and co-curricular projects. An example of such an extended investigation can be built on the risk problem mentioned under Section 3 above. Students can compute different risks from the information given in the report and other sources, discuss the science of food, compare types of healthy diets from different ethnic groups, and write about their personal experiences of committing to healthy living. This type of learning experience develops cooperative, social, and communication skills, motivation to contribute a fair share of effort to achieve team success, and problem solving. Team leaders can develop leadership and organizational skills, and every member should be given the opportunities to develop these 21CC skills.

7 21CC Mathematics Teachers

To be effective in teaching 21CC to their students, mathematics teachers in the 21st century must have developed attributes similar to these four types of outcomes so that they can model the processes required to achieve these outcomes. This should be a life-long learning goal for the teachers.

Confident teachers.
They must be confident in their mathematics content knowledge and pedagogical content knowledge so that they can make informed decisions about how best to implement their lessons. They must be digitally literate in using ICT tools to provide enriched learning experiences that match the digital experiences of their students. The confidence they display in front of the class can inspire their students to work hard on the assigned tasks.

Self-directed learners.
They must take full ownership of their professional development, instead of depending on assigned mentorship and recommendations of their school's staff developers. They will benefit from a mindset that treats every lesson as a miniature experiment to learn more about how their teaching might impact student learning cognitively, metacognitively, and emotionally. As the curriculum is likely to be changed periodically, they must be proactive in learning on their own new mathematics and pedagogy required by the revised curriculum. As per student learning, teachers must also learn to ask mathematics questions as well as education-related questions. In the 21st century, teachers as well as students can tap on international experts through many forms of online connections to enhance their learning, for example, by taking courses from Coursera (http://www.coursera.org).

Concerned citizens.
They must keep abreast of current and significant local and global issues and share their concerns about these issues with their students. They can show how mathematics can be helpful in understanding these issues in a numerate and rational way.

Active contributors.
They may volunteer to organise activities to address issues of particular concern to them. They will apply their mathematical knowledge or learn

new mathematics to make their contributions really count. Finally, they can share their professional development journeys with the community of mathematics educators and through social media.

8 Concluding Remarks

The above sections show how the widely circulated 21st century competencies can be encapsulated by four types of student outcomes as delineated by the Singapore 21CC framework. In short, confidence in applying one's mathematics knowledge and reasoning can be cultivated through exercises in critical thinking, problem posing, and research. This confidence is the basis of self-directed learning when the students enrich their learning through four types of metacognitive attributes. Teachers can arouse concerns passions of their students about current issues by getting them to work on mathematics applications and modelling, and encourage them to make active contributions individually or in teams to resolve some of these concerns. These four outcomes do not necessarily proceed in the linear fashion as indicated, although this progression is a logical way to make connections among them. The optimal progression will have to wait for new research into this area.

In an entertaining book consisting of mathematics stories and puzzles, Constanda (2009) claimed that "In mathematics … you must always be prepared for the unexpected" (p. 268). Likewise for students in this 21st century, learning and life events are full of surprises and uncertainty, and mathematics can be a powerful tool to understand and deal with events in this VUCA world: *Volatility, Uncertainty, Complexity,* and *Ambiguity*. Mathematics teachers, individually and collectively, must capitalise on every opportunity, however minor, to continually infuse 21CC into their mathematics lessons. Future generations of 21st century students deserve this commitment from their teachers. The rewards for both teachers and students in this joint effort will be tremendous and mutually satisfying, indeed.

References

Berger, W. (2014). *A more beautiful question: The power of inquiry to spark breakthrough ideas*. New York, NY: Bloomsbury.

Brafman, O., & Brafman, R. (2008). *Sway: The irresistible pull of irrational behavior*. New York, NY: Doubleday.

Brown, S.I., & Walter, M.I. (2005). *The art of problem posing* (3rd ed.). Mahwah, NJ: Lawrence Erlbaum Associates.

Buoncristiani, A.M., & Buoncristiani, P. (2012). *Developing mindful students, skillful thinkers, thoughtful schools*. Thousand Oaks, CA: Corwin.

Chua, P.H. (2011). *Characteristics of problem posing of Grade 9 students on geometric tasks*. Unpublished PhD thesis, Nanyang Technological University, Singapore.

Constanda, C. (2009). *Dude, can you count? Stories, challenges, and adventures in mathematics*. London: Springer.

Efklides, A. (2006). Metacognition, affect, and conceptual difficulty. In J.H.F. Meyer, & R. Land (Eds.), *Overcoming barriers to student understanding: Threshold concepts and troublesome knowledge* (pp. 48-69). London: Routledge.

Flavell, J. H. (1976). Metacognitive aspects of problem solving. In L. B. Resnick (Ed.), *The nature of intelligence* (pp. 231-235). Hillsdale, NJ: Lawrence Erlbaum.

Hanushek, E.A., & Woessmann, L. (2015). *Universal basic skills: What countries stand to gain*. Paris: OECD Publishing.

Hattie, J.A.C. (2009). *Visible learning: A synthesis of over 800 meta-analyses relating to achievement*. London: Routledge.

Karpicke, J.D., Butler, A.C., & Roediger III, H.L. (2009). Metacognitive strategies in student learning: Do students practise retrieval when they study on their own? *Memory, 17*(4), p. 471-479.

Kaur, B. (2011). Mathematics homework: A study of three grade eight classrooms in Singapore. *International Journal of Science and Mathematics Education, 9*(1), 187-206.

Lee, K.Y. (2013). *The wit & wisdom of Lee Kuan Yew* (1923 – 2015). Singapore: Edm. [Reprinted in 2015]

Lee, N.H., & Ng, K.E.D. (Eds.). (2015). *Mathematical modelling: From theory to practice*. Singapore: World Scientific.

Mason, J., Burton, L., & Stacey, K. (2010). *Thinking mathematically* (2nd ed.). Harlow: Pearson.

Ministry of Education, Singapore (2010). *MOE to enhance learning of 21st century competencies and strengthen art, music and physical education*. Retrieved 31 December, 2015 from www.moe.gov.sg

National Research Council. (2005). *How students learn: Mathematics in the classroom.* Committee on How People Learn. Washington, DC: National Academy Press.

Nelson, D., Joseph, G.G., & Williams, J. (1993). *Multicultural mathematics: Teaching mathematics from a global perspective*. Oxford: Oxford University Press.

Oakley, B. (2014). *A mind for numbers: How to excel at math and science (even if you flunked algebra)*. New York, NY: Jeremy P. Tarcher/Penguin.

Ooten, C. (with Moore, K.). (2010). *Managing the mean math blues: Math study skills for student success* (2nd ed.). Upper Saddle River, NJ: Pearson Education.

Partnership for 21st Century Learning. (n.d.). Retrieved 12 January, 2016 from http://www.p21.org/

Pang, M. (2012, July 14). Fast food 'can be life-threatening'. *The Straits Times.* Retrieved 12 January, 2016 from https://www.healthxchange.com.sg/News/Pages/Fast-Food-Can-Be-Life-Threatening.aspx

Rohrer, D., Dedrick, R.F., & Burgess, K. (2014). The benefit of interleaved mathematics practice is not limited to superficially similar kinds of problems. *Psychonomic Bulletin & Review, 21*(5), 1323-1330.

Schraw, G., & Moshman, D. (1995). Metacognitive theories. *Educational Psychology Review, 7*(4), 351-371.

Sinking of MV Sewol. (n.d.). In *Wikipedia.* Retrieved 12 January, 2016 from https://en.wikipedia.org/wiki/Sinking_of_MV_Sewol

Teng, A. (2015, May 14). Singapore tops world's most comprehensive education rankings. *The Straits Times*.

Ultimatum Game. (n.d.). In *Wikipedia*. Retrieved 12 January, 2016 from https://en.wikipedia.org/wiki/Ultimatum_game

Wiliam, D. (2011). *Embedded formative assessment*. Bloomington, IN: Solution Tree Press.

Wong, K.Y. (2003). Mathematics-based national education: A framework for instruction. In K.S.S. Tan, & C.B. Goh (Eds.), *Securing our future: Sourcebook for infusing National Education into the primary school curriculum* (pp. 117-130). Singapore: Pearson Education Asia.

Wong, K.Y. (2013). Metacognitive reflection at secondary level. In B. Kaur (Ed.), *Nurturing reflective learners in mathematics: Association of Mathematics Educators Yearbook 2013* (pp. 81-102). Singapore: World Scientific.

Wong, K.Y. (2015a). *Effective mathematics lessons through an eclectic Singapore approach: Yearbook 2015, Association of Mathematics Educators*. Singapore: World Scientific.

Wong, K.Y. (2015b). Use of student mathematics questioning to promote active learning and metacognition. In S.J. Cho (Ed.), *Selected regular lectures from the 12th International Congress on Mathematical Education* (pp. 877-895). Cham, Switzerland: Springer.

Wong, K.Y., & Tiong, J. (2006). *Developing the Repertoire of Heuristics for Mathematical Problem Solving: Student problem solving exercises and attitude* (Unpublished technical report). Singapore: Centre for Research in Pedagogy and Practice, National Institute of Education, Nanyang Technological University.

Wong, K.Y., Zhao, D.S., Cheang, W.K., Teo, K.M., Lee, P.Y., Yen, Y.P., Fan, L.H., Teo, B.C., Quek, K.S., & So, H.J. (2012). *Real-life mathematics tasks: A Singapore experience*. Singapore: Centre for Research in Pedagogy and Practice, National Institute of Education, Nanyang Technological University.

Chapter 4

Mathematics in 21st Century Life

Barry KISSANE

School mathematics in the 21st century seems likely to be characterized in part by new expectations regarding the complex lives of students after they leave school. In this chapter, an analysis of four domains of 21st century life is provided: working, shopping, participating in society and personal satisfaction. A productive worker will need to use the mathematics they have learned at school in the increasingly rapidly changing world of work. A careful consumer will need to develop expertise in understanding the many choices they need to make regularly over the course of their adult lives. The informed citizen will need to engage with, understand and even contribute to the rapidly changing world of information in the 21st century for both personal and national benefit. The balanced person will need to become aware of and appreciate mathematical perspectives on everyday events and on mathematics itself. Some examples are offered in each of these domains, recognizing that an important part of the craft of mathematics teaching is to interpret and implement the curriculum to achieve the best outcomes for students. While some of these outcomes are determined by short-term transition from school to later study, others require a longer perspective.

1 Introduction

Today's secondary school children in Singapore are likely to be adults from the present time until around 2080 or so, a period which will comprise about two thirds of the 21st century. In this chapter, we explore the several dimensions of that adult life, to enquire about the extent to which the mathematics curriculum will meet their needs.

For most people, the dimensions of adult life are related to a variety of roles that people undertake, sometimes simultaneously. These roles include the following at various stages:

- The successful student
- The productive worker
- The careful consumer
- The informed citizen
- The balanced person.

The first of these roles is fairly widely understood, and hence will not be a primary focus of this chapter. It is well recognized by students, their teachers and parents that competent study of mathematics is needed in school in order to achieve goals after school, and decisions about course content are often decided externally to the student and the teacher. For example, future scientists, engineers, psychologists, doctors, economists and other professionals will study the mathematics deemed necessary by their professions, following stipulated courses beyond school.

Accordingly, this chapter will focus attention on the remaining four roles, which are usually less explicitly identified in school mathematics, but which are nonetheless of importance for all students (including those who embark on further study of mathematics).

2 The Productive Worker

Today's students will spend many years engaged in productive work, which will require them to make use of the mathematics they have learned at school. For some time now, school curricula have tried to emphasize the practical applications of mathematics in the everyday world, including the workplace. This is a difficult task, in part because adolescent students have trouble seeing themselves in the workplaces concerned, and in part because it is difficult to know in advance what mathematical thinking is actually involved in working.

A recent project in Australia investigated the place of mathematics in work in an interesting way. The project involved the Australian Association of Mathematics Teachers (AAMT) and the Australian Industry Group (AIG) (2014a) in a collaboration involving case studies of entry-level workers in various fields. Teacher-researchers studied a variety of workplaces to understand the significance of mathematics for work, and to compare the expectations on new workers with the school mathematics curriculum. The workers concerned were not university graduates, but were non-specialists in mathematics, so that the study provides insights into the role of mathematics for those who are not quantitative specialists. A major outcome of the research was a *Quantitative Skills Map*, outlining the ways in which mathematics was of significance in a wide range of fields, which included engineering, drafting, retail, manufacturing, mining and national defence.

The map (AAMT & AIG, 2014a) included four key elements, which are described in some detail in the report:

- Mathematical Content
- Mathematical Level
- Mathematical Executive Functions
- Workplaces as Technologically Rich Environments.

Of particular interest are the Mathematical Executive Functions identified, which highlighted the differences between mathematics at school and mathematics in the workplace. Workers needed a blended set of skills and capabilities that involved understanding mathematical concepts, understanding the practical tasks they confronted in the workplace and understanding the strategic processes of bringing these together in productive ways (AAMT & AIG, 2014a, p. 2). Mathematics in the workplace was identified as different in many respects from mathematics in school, and the report identified several aspects of a "mismatch between young peoples' mathematics skills and the expectations of modern workplaces." (p. 42)

The study outlined the complex tasks in which workers needed to use their mathematics, by referring to the executive functions:

Workers use executive functions to perform activities such as planning, organizing, strategizing, paying attention to and remembering details and managing time and space. The three components of the executive functions included:

- Resisting impulses – stopping to allow time to think before taking action;
- Effective use of working memory – bearing a number of things in mind while thinking about how to do something; and
- Cognitive flexibility – a capacity to think about a problem in different ways. (AAMT & AIG, 2014b, p. 3)

Research of this kind offers some insights into the difference between how students learn and use mathematics in school and how they will later make use of their mathematics in the world of work. The difference is less about the content of the mathematics than it is about the circumstances in which the mathematics is used in practice. While it is inevitably difficult for school mathematics to provide contexts that are regarded as realistic by students, there would seem to be a place for a closer link between school and work for many students in their final years of school. An improved understanding of the nature of mathematics in workplaces would help both teachers and students to make the transition manageable, but would also support further developments in the school curriculum to address this purpose.

An additional observation of this study was the increasing importance of technology of many different kinds in workplaces. The use of spreadsheets and graphical outputs was commonly observed, but other technologies have become a standard part of the modern workplace. This suggests that there is a need to embed digital technologies into the mathematics curriculum so that they are not seen as optional tools, but a standard feature of the quantitative environment in which people will work throughout their working lives. It was clear from the case studies that technology is changing rapidly the nature of work and the nature of the mathematics that is used. If schools are to prepare students for lifelong careers as productive workers, there would seem to be a need to consider more carefully how technology and mathematics might be connected at school, as well as in the workplace. These studies have demonstrated one way to acquire and to disseminate such information.

Unsurprisingly, teachers have few opportunities to engage with modern workplaces, so it seems important to find ways to provide teachers with better and up-to-date information about the nature of modern work, in order to help them prepare students to make a smooth transition from school to work.

Other studies have also examined the mathematical needs of workers beyond school. The Organisation for Economic Co-operation and Development (OECD) has recently begun supplementing the well-known PISA studies with studies of adult competencies. A recent OECD initiative, the Programme for the International Assessment of Adult Competencies (PIAAC) has begun to examine:

> How skills are used at home, in the workplace and in the community; how these skills are developed, maintained and lost over a lifetime; and how these skills are related to labour market participation, income, health, and social and political engagement. (OECD, 2013).

The PIACC skills, not just restricted to mathematical skills, include a range of integrated skills such as interpreting information and engaging with technology in various ways. The initial PIACC work has thus explored the literacy, numeracy and problem-solving in technological environments of adults. Although work in this area is relatively new, it might also be expected to offer insights into how to improve the connections between the world of school mathematics and the adult world of work and other activities, in addition to offering insights into good ways of building on school experiences for workplace training. While important, mathematics is not the only element in this work; building strong connections among mathematics, technology and language in suitable contexts will be an important element of productive workers in the 21st century. It is more challenging to obtain information about adults than about school children (since they are not conveniently accessible via schools), but it is hoped that the imminent PIACC reports about Singapore in particular will be informative to teachers and others.

3 The Careful Consumer

Regardless of their occupation, students of today will become the consumers of tomorrow, responsible for making good decisions on the everyday tasks of living in a modern world. Many of these decisions require a level of mathematical understanding that is not always recognized and not always focused upon in school. These decisions affect many parts of the lives of families, as well as the economy that includes them. For example, consumers make many of their own decisions regarding their housing, health, food, transportation, vacations, children's upbringing and leisure time. Mathematics is involved in most of these decisions, although frequently this is not recognized. In this section, we look briefly at some illustrative examples.

3.1 *Shopping*

The everyday task of grocery shopping in a modern supermarket often involves mathematical thinking. Consumers need to choose among a variety of products, with prices and quantities varying, differing ingredients, discounts provided in various ways and often in the midst of advertising pressures. Of course, habit can prevail, with shoppers routinely choosing again the same products as they have chosen previously, although mathematical thinking can also be brought to bear on such matters. To illustrate the situation with a typical example, the author recently needed to replenish his supply of sweeteners, used in his daily coffee. The sweetener label is shown in Figure 1.

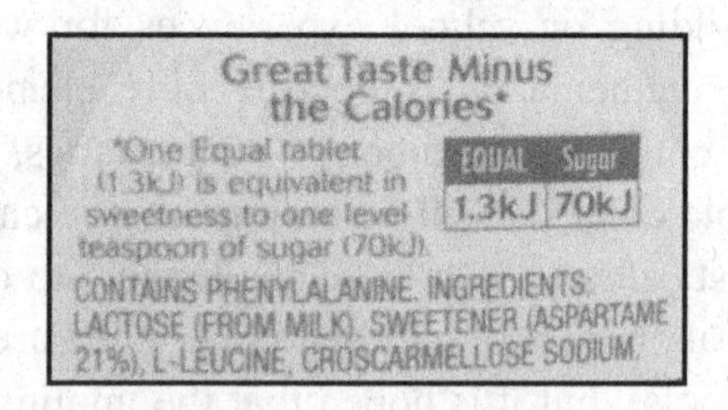

Figure 1. Sweetener label

On arriving at the supermarket, a bewildering array of alternatives is available, some of which are shown in Figure 2:

Figure 2. Some alternative sweeteners

On closer inspection, the choice is not merely one of preferences, according to individual tastes. Different quantities of sweeteners are available, for different prices. Some use different ingredients. Figure 3 shows some illustrations of these differences.

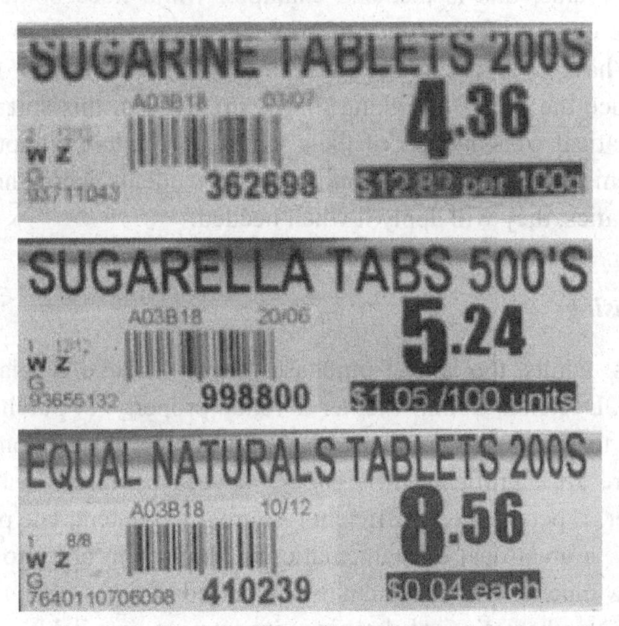

Figure 3. Comparing sweeteners

The supermarket labels displayed, in accordance with Australian consumer laws, provide some information to assist shoppers with making comparisons among products, but in this case one unit price is displayed per 100g, another is displayed per 100 units (tablets) and a third displayed per single tablet, so that the consumer still needs significant mental arithmetic skills to compare these from a financial point of view.

Compounding these comparisons, some sweeteners are twice as sweet as others, so that tablets need to be broken in half to obtain an equivalent amount of sweetness. While the obvious comparison is about price, some consumers might seek to compare details of ingredients for some purpose, such as health or taste preferences, requiring a careful mathematical comparison of the fine details of products. In addition, decisions of this kind often need to be made in a brief period of time, and without the benefit of either a calculator or pen and paper, requiring confidence and fluency with mental estimations in order to reach optimal decisions.

Of course, this is just one example, while householders typically purchase many different kinds of products. As Kemp (2012) and Kissane (2012) have suggested, it would seem important for students to experience the kind of thinking that is involved in this sort of informal mathematical work as part of their school curriculum, although there is an optimistic tendency to assume that if students learn the necessary mathematics, they will apply it when needed.

3.2 *Housing*

For most adults, the largest purchase they will make is related to their housing. Decisions about whether to rent a property, or purchase a flat or a house to live in are not available to all people, for economic reasons, and there are significant differences between countries in this respect. However, if people have sufficient resources to contemplate purchasing a property, a great deal of mathematics is required in order to understand well how much it will cost them over a period of time.

The mathematics of housing investment inevitably requires an understanding of compound interest when regular repayments are made, so that, effectively, a form of reducible interest is involved, since the amount of mortgage repayment that is effectively paying interest on a loan will change over time. The situation becomes even more complex if the changing value of money is taken into account (in the form of some kind of inflation).

The school curriculum generally helps students to understand simple interest, which is almost never used in practice (even though the basic principle of using an interest rate is employed to model other kinds of financial situations). While students often study compound interest, they generally do so in the context of depositing money into an account, and leaving it to accumulate interest (a rare practice for most people). To handle the mathematics of reducible interest, and thus to understand their housing finance, students in fact need to use a suitable technology to handle the calculations. A spreadsheet on a computer or a calculator is sufficient for this, but it is rare for this to be part of the standard school curriculum.

If the details of the mathematics are clear to people, they are then enabled to consider the effects of varying the parameters of their housing purchase, such as starting with a larger deposit, changing the duration of the loan, making extra cash deposits or finding opportunities with different interest rates. Without a means of undertaking these kinds of investigations, consumers are left entirely at the mercy of the people selling the properties to them, which seems undesirable. Again, the school curriculum and the teacher in the classroom might consider the practicalities of computing interest, in addition to a study of simple and compound interest, in order to address the needs of adult consumers in the 21st century.

3.3 *Health*

Another area in which consumers require some mathematical thinking concerns their health. Kemp (2012) provided several examples of suitable activities related to health, especially to diet and its consequences, using everyday materials such as the labeling on food packets. Adults in the 21st century will need to have both the experience and the inclination to attend to information of these kinds in order to develop and maintain a healthy lifestyle.

People are not always healthy, however, and in the modern world, testing for medical conditions has become common and is often recommended as a precaution when symptoms suggest that there might be problems. Typical examples are screening tests for breast cancer for women and prostate cancer for men. However, no tests are perfect, and it is important that consumers have sufficient mathematical competence to understand the resulting information, in conjunction with the advice from medical practitioners. In fact, Gigerenzer (2002) has undertaken considerable research that suggests that many medical practitioners themselves struggle with the mathematics involved, making it even more important that consumers themselves are able to understand the significance of results.

To explore an example, the case of breast cancer is well described on the *Understanding Uncertainty* website in the UK (2008). The website describes mammography tests among 50-70 year old women in the UK, noting three important aspects:

- Mammography detects approximately 85% of breast cancers.
- Around 10% of women with no cancer will get a positive result (in error).
- About 1% of the population age group has breast cancer.

If a positive test result is returned following a mammography, how likely is it that the woman concerned actually has breast cancer (rather than an erroneous test result)? It seems from Gigerenzer's studies, and those of other people, that this is a surprisingly difficult question to answer, and it is not unusual for people to over-interpret the risk, sometimes with undesirable consequences (such as invasive surgery or treatments), as well as substantial worry for patients themselves and their families. This problem is well-known amongst medical people and educated people, but less well-known by the general public.

It can be argued (as both Gigerenzer and the website suggest) that the problem is better understood if percentages (which are in effect, conditional probabilities) are avoided. Thus, natural numbers are used in

Table 1, relying upon the percentages to show the expected results of 1000 women undergoing the mammography test.

Table 1
Representing mammography tests in natural numbers

	Test is positive	**Test is negative**	**Total**
Cancer	9	1	10
No cancer	99	891	990
Total	108	892	1000

From this way of thinking about the test, it is clear that only about 9 of the 108 positive test results are returned by women with cancer. In other words, a positive test result suggests about 1 chance in 12 or 8% probability that a woman has breast cancer. Many people interpret the (fallible) test much more severely than this. Indeed, Blastland and Spiegelhalter (2013, p. 263) cite a UK study reporting that three times as many women are treated unnecessarily for every woman whose life is saved by screening for breast cancer.

Of course, tests may be more reliable in the future, and other diagnostic information is used as well, but it seems important that health consumers be empowered by mathematics to be able to interpret information provided to them to some extent, or at least to be sufficiently aware to ask their doctors important questions, rather than accepting results uncritically. In this case, the (difficult) mathematics of conditional probability might be involved, although – as noted – it may be more insightful to use numbers than percentages. It may be optimistic for schools to teach students conditional probability and to assume that they are then able to use it when needed; it seems instead more realistic for work of this kind to be included in the school mathematics curriculum.

There are many everyday matters related to health that also involve adults making careful personal decisions, requiring a level of mathematical competence to interpret the information presented to them by health officials or accessible from other sources. Examples of these explored in various ways by Blastland and Spiegelhalter (2013) include vaccination, contraception, taking drugs, alcohol, tobacco, childbirth and surgery.

3.4 *Gambling and intuition*

Gambling remains a popular pastime in many countries, including both Singapore and Australia, and it seems likely to continue to be a feature of life in the 21st century. A good recent discussion by Blastland and Spiegelhalter (2013, pp. 129-138) observes that some modern forms of gambling – such as lotteries – are sometimes not even recognized as gambling, while other forms of gambling (such as internet gambling) are becoming increasingly popular, and problem gambling by individuals is now recognized as a significant problem. A careful consumer needs to develop some appreciation of how gambling works to help avoid personal problems associated with gambling; accordingly it seems rare, anecdotally at least, for mathematicians and statisticians to engage in excessive gambling, as they are more likely to understand the risks and likely outcomes of many legal forms of gambling. Although it is not uncommon for the school curriculum to refer to elementary probability, it is less common for the mathematics of gambling to be addressed explicitly in school, in part because gambling is regarded as an 'adult' activity and in part because it is difficult for schools to teach about gambling without parents and others misinterpreting this as encouraging students to gamble.

Without a good mathematical understanding of probability theory and expectations associated with gambling, which often require relatively sophisticated mathematics, people are understandably inclined to rely on their intuitions, when confronted with gambling opportunities. Perhaps, then, a minimum curriculum in mathematics might direct attention to helping students realize that their intuitions are frequently incorrect: that some events happen more often than they expect, while others are rarer than they might expect.

The traditional 'birthday problem' is a good example, and easily used in a classroom. The problem concerns whether or not two people in the same room share a common birthday (such as 23 April, irrespective of the year of birth). Because there are many possible birthdays (365 or even 366 in leap years), most people regard it as surprising to find someone with the same birthday as themselves, and so a common intuition is that a small group of people (such as might be in a classroom or a pair of football teams) is unlikely to have at least two people with the same birthday. It is quick and easy to test this intuition in practice in a classroom and many students will be surprised to learn that their intuition is defective. Since an actual class can be used only once, a good adaptation of this problem using a spreadsheet is provided by Tay (2011). A probabilistic analysis, suitable for older students, is available at Wikipedia (Birthday Problem, n.d.), and shows that a group of only 23 people is sufficient for the probability of match to be about 50%, while a group of 40 random people has a greater than 90% chance of two people sharing a birthday.

Students might also have opportunities to see the dangers of paying too much heed to our intuitions from other probabilistic activities that will test them. Good examples are the Cereal Box problem and the Monty Hall problem, both well-known and described well with suitable online simulations (e.g., Baetz & Reese, 2010; Utah State University, n.d.). Activities like these can also be studied formally by older students, but there is value in students seeing that their intuitions, and those of other people, can be quite wrong, a valuable and informal lesson to learn when they later confront gambling situations.

Other classroom activities may have a similar character, to allow students to see that their intuitions are unreliable and perhaps later to be studied by some students in more advanced classes. For example, if students in a class are asked to write down what they regard as a typical sequence of ten random tosses of a fair coin (such as HHTTHTHTTH, recording successive results as tails or heads), the class intuitions regarding what is 'typical' can be easily tested. If the intuitive results are compared with theoretical results, many students in most classes will be surprised:

- It is more likely that four or six heads will occur than five heads.
- About half the time (in ten tosses) there will be a run of at least three heads in a row.

- About 80% of the time, there will be a run of at least three heads or at least three tails in a row.

Most students are surprised to find that their idea of 'typical' is often incorrect, a valuable lesson for later confronting situations involving gambling.

3.5 *Insurance*

Another feature of modern life, likely to persist into the 21st century, is insurance. Consumers need to decide which kinds of insurance, if any, to purchase and to understand some of the mathematics involved in order to make good decisions. The mathematics of insurance is unfortunately quite complex, and is generally undertaken by actuaries in large and powerful insurance companies, so that it is not reasonable for the school curriculum to attempt to deal with the details. However, consumers need to decide about the merits of life insurance, house insurance, medical insurance, vehicle insurance, travel insurance, etc. and to make comparisons between different products, so that it seems important that the mathematical ideas are at least introduced at some stage of school.

The most likely place for this is in courses concerned with probability and statistics, where students might learn about the broad mechanisms used by insurance companies to determine risks and to evaluate expectations. The links between data (such as life tables and accident and crime statistics) and the associated empirically derived probabilities can be made clear to students. Similarly, even if the theoretical concept of expectation is too difficult for many students, a reasonable understanding of the consequences of insurance can be obtained via computer simulations on spreadsheets, computer software or calculators.

4 The Informed Citizen

To succeed and to prosper, modern democracies like those in Singapore and Australia rely on developing and maintaining an informed citizenry.

Informed citizens are aware of current issues of the day and have a capacity to engage with them personally to reach their own opinions, rather than relying entirely on the views of others, such as friends, journalists or politicians. A democracy is strengthened by having an educated population willing and capable to analyze situations to reach strong conclusions and – ultimately – vote accordingly to ensure that the best decisions are reached. In this context, it is worth noting that the PIACC work, as described earlier, has identified the importance of social and political engagement by adults. (OECD, 2013).

As the early years of the 21st century have already made clear, citizens will live in an information age, and mathematics will be an important element in their effective engagement with information. Information is available in many forms, including newspapers, magazines, online sites and social media. Information is often compressed for efficient display into tables of data or graphs or various kinds, as well as text and speech. Each of these forms requires mathematical expertise to deal with.

Previous AME *Yearbooks* have provided extensive advice and examples of how people might use mathematics for these purposes, so that they are reaching their own conclusions to some extent, rather than relying on the interpretations and conclusions of others. Kemp (2010) described and exemplified her Five Step Framework for interpreting graphs and tables, and gave several Singaporean examples of the framework in use. The framework provides systematic advice and specific help to students to unpack the data that have been compressed into tables and graphs. She noted that research has indicated that analysis of information is often more complex than is realized, and students need to develop expertise in distinguishing sound interpretations from others. Thus, people need to think carefully about the arguments presented by the media and the evidence that is given to support them. They also need to be aware that there are often defects in the graphs and tables that can unintentionally, or intentionally, mislead or misinform the reader. (Kemp, 2010, p. 200)

Similarly, Kissane (2012) described and illustrated several aspects of 'numeracy', concerned with using mathematical ideas and mathematical thinking to understand and interpret everyday situations. Many everyday tasks, including those related to information presented in various media, require students to make use of the mathematics they have already learned in novel ways. He described a number of Australian projects that have concluded that numerate behavior does not necessarily happen without conscious attention in the school curriculum. If we want numerate citizens, it is optimistic to expect that learning mathematics by itself will be sufficient to realize that goal.

A characteristic of public information in recent years is the widespread use of digital media, in addition to the traditional printed media of newspapers, magazines, radio and television. The informed citizen will be expected to use their mathematical expertise in a range of settings, many of which will not have been the explicit focus of schooling. Thus, for example, the Singapore Department of Statistics provides a facility for online users to construct their own tables from the available data (using a *Table Builder* facility) rather than relying on tables constructed by Departmental staff.

To elaborate and improve upon this concept, the Department now publishes the *SingStat Mobile App* (2015), in addition to conventional data presentations in published books and online possibilities, so that users can access data of interest to them whenever they wish and wherever they are. The example in Figure 4 shows that users of the app can generate their own information to answer questions of their own interest, which of course still requires expertise to interpret the information competently.

In this case, a user can see from the graph that, while the number of mobile phones in Singapore continued to rise steadily from 2007 to 2013, the number of SMS text messages sent after 2011 has declined steadily and significantly, perhaps suggesting that people are communicating with each other in different ways from SMS and even that the mobile phones recently are likely to be smartphones. Interpreting the graph is made a little more difficult by the use of two different vertical scales (one for millions and the other for billions), which of course is accommodated in Kemp's framework (2010).

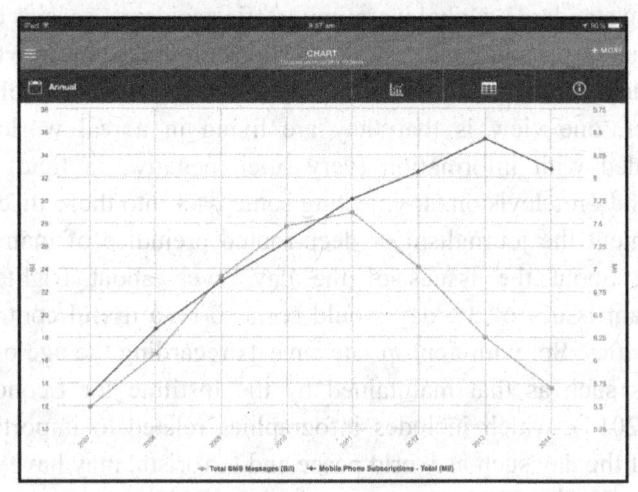

Figure 4. SingStat graph comparing SMS and phone use

Online presentations of information are already changing, less than two decades into the 21st century, so it seems reasonable that they will continue to change in ways that are not yet clear. An example of a recent change is for information to be presented online in a way that allows the user to interact with it, as part of the task of interpreting it. For example, Jericho (2015) describes recent changes in the Australian economy for public readership, including a series of graphs that can themselves be manipulated in various ways by the reader. The manipulations include an ability for readers to select different variables or time periods for analysis, or even to obtain the data being represented graphically, from original sources, should they wish. Active reading of this kind will require new mathematical skills to be acquired and exercised by citizens to become adequately informed about the events and issues of the day. Tools for people to present their own data online are now available, and it seems important for mathematics teachers to stay abreast of these kinds of opportunities, which seem likely to become more important into the century, and to help their students explore their productive use. A good example is the open source software for journalists and others, *Datawrapper* (2015), which was used by Jericho (2015).

Statistics seem to be often collected and publicly disseminated on bad news, rather than good news. (Causes of death, disasters, accidents, declining standards, global warming, traffic accidents, species extinction, etc.). This raises some interesting questions on the extent to which it is a good idea to focus attention on these things with school children. However, one view is that they are living in a real world and, are bombarded with information (very often negative in tone) on social media and on television, so injecting some data into those discussions to complement the journalism or deep-rooted prejudice of many kinds in thinking about the issues of the day, even about frightening and unpleasant issues of the day would seem to be a useful contribution of mathematics. So, with appropriate caveats regarding the age of children, websites such as that maintained by the Institute for Economics and Peace (2015), which includes infographics related to important global issues of the day such as world peace and terrorism, may have a valuable place in school mathematics.

In a wider domain, in recent years the presentation of public information online has undergone a transformation, in which people can construct their own representations of reality to convey a message or tell a story for others to use. The infographics on the Institute for Economics and Peace (2015) website are good examples. Perhaps the best example of this at present uses the *Tableau Public* software (2015) which is used to highlight a data visualization each day (called *Viz of the Day*). The visualizations can be downloaded by users, who can then interact with them, unlike the case of static information in a table or a graph or in text. Citizens in the 21st century can be expected not only to engage with and interpret such visualizations, but also to become producers of information to inform others. Some of the skills involved are certainly mathematical, and students have a right to expect that their schooling will equip them to undertake activities of these kinds into their adult lives. Again, a mathematics curriculum for the 21st century might consider how to encourage and support students to do this.

As well as representing data, news and other media such as the Internet are often used to inform the public about important breakthroughs in science and technology, although these are sometimes selectively reported in sensationalist ways, to attract more readers or viewers to a news item. A good example of this occurred in October 2015 following the publication of a World Health Organization press release (International Association for Research on Cancer, 2015) at the same time as a research report was published in the prestigious *Lancet* journal in the UK. The press release noted, "The experts concluded that each 50 gram portion of processed meat eaten daily increases the risk of colorectal cancer by 18%" (p. 1). The same study also reported an increased risk of 17% associated with eating 100 gram of red meat per day.

News media in many countries use such professional information as a source of news, which in a number of cases was reported with an alarmist headline, such as "Bacon, sausages, ham and other processed meats are cancer-causing, red meat probably is too: WHO" (Australian Broadcasting Commission, 2015). Such headlines generated a level of fear in the public and a corresponding level of hostility by meat industries. Informed citizens might be expected to be careful readers of media of these kinds, and realize that the actual risks of contracting colorectal (CR) cancer were not published, only the extent to which they were increased by meat consumption. Few of the media reports clarified the precise meaning of increasing a risk by 17% or 18%, and it seems likely that many readers naively interpreted the reports as suggesting that these were in fact the probabilities of meat eaters contracting the cancer.

In an Internet age, informed citizens might be expected to both realize that the cancer risks were not published and to seek them online to understand well the levels of personal risk involved. For example, US data provided by Centers for Disease Control and Prevention (2015) indicate that the probability of a (US) woman aged 50 contracting CR cancer in the next ten years is 0.0052, in the next 20 years is 0.0137 and the next 30 years is 0.0269. Put another way, about 1.4 US women in a hundred will develop this cancer sometime between 50 and 70 years of age. So, about 98.6 women in every hundred will *not* do so between 50 and 70. While cancer is an awful disease, these seem to be modest risks.

If a woman consumes 100 gram of red meat a day, however, according to the press release these risks will rise by 17% to, respectively, 0.0061, 0.0160 and 0.0315; in that case, about 1.6 women in 100 will develop the cancer between ages 50 and 70. While that's certainly an increase, it is questionable whether such an increase ought to have caused so much apparent international alarm. Indeed, the reaction in several countries makes clear that many citizens do not (or are not able to) read news media with due care, and are easily influenced by alarmist headlines.

Informed citizens can be expected to look beyond the headlines and use mathematical thinking to analyze carefully what they read. In this case, some online discussions such as Harcombe (2015) challenging the interpretation and the presentation of the results of the research were quickly available online. Indeed, possibly as a result of the reactions to the extensive media reports (not to the research itself), the World Health Organization (2015) also published further clarifying information on the matter. To interpret carefully information presented in the media, tomorrow's informed citizens need to be educated today to engage in this kind of activity, and the school mathematics curriculum has an important role in helping them to do so.

5 The Balanced Person

The final contribution of mathematics to the lives of 21st century citizens is concerned not with the use of mathematics, but with its appreciation. In modern affluent societies like those of Singapore and Australia, lives can be enriched through the use of leisure time through mathematics. While leisure time is a relatively new concept, people of the 21st century can be expected to live longer and work less than those of earlier times, so that an opportunity to enrich their lives through access to mathematics exists. Mathematics as a source of pleasure and enrichment is no less deserving of a place in the world of 21st century people than literature, media, the arts and sport, but is commonly neglected. A balance between sources of personal pleasure that are intellectual, aesthetic and physical is desirable in a society that is fortunate enough to support all of these.

There would seem to be two main sources of written mathematical materials for people to read for pleasure: printed materials and online materials. As noted elsewhere (Kissane, 2009), there is now a substantial body of mathematical writing available for popular consumption, although it seems not to be widely known. In Australia, the best bookshops often have mathematical materials listed (incorrectly) in a category of 'popular science', which may contribute to the problem. The best bookshops in Singapore have many examples, modern and otherwise, of mathematical writing intended for a wide audience.

Introducing students to such literature can be quite taxing, however, especially if students do not realize that this species of literature exists and can be most enjoyable to read. School libraries are sometimes focused sharply on the direct needs of school courses, such as textbooks and problem books, and it is not unusual to find school libraries with very few mathematical materials in them intended for reading. Yet if students are not introduced to the literature about mathematics when at school, they are unlikely to engage in reading it after they leave school: unlike literature and even science, it is very unusual for students to receive gifts of mathematics books; there are very few publicly available magazines about mathematics and it is unusual to see mathematical writing in newspapers or mathematical argument and representations in film and television. So the only opportunity they have to find out about the world of written mathematics and mathematical argument may be whilst at school. A systematic way of encouraging students to read is worthy of being sought, and may have life-long benefits.

An exception to these discouraging observations is that it is possible to do better. As Merow (2015) described, the Argentinian mathematics professor and journalist, Adrián Paenza was recently awarded a prestigious prize from the International Mathematical Union for his work in popularizing mathematics in Latin America and Europe through newspapers, television programs and children's books. The newly established Leelavati Prize for mathematical outreach is awarded every four years from 2010, recognizing at the highest levels the importance of mathematics being provided in an appealing form for a wide audience.

Perhaps ironically, the online world may be a better source for students to encounter pleasurable reading, at least at first. There are a number of free online magazines, columns and blogs related to mathematics, some of which will appeal to school students, while others are intended for a more sophisticated audience, although not necessarily an audience with professional qualifications in mathematics. Many of these are collated for convenience by Kissane (2015), and include regular mathematical columns sponsored by the Mathematical Association of

America, mathematical blogs gathered by the American Mathematical Society, a poster series entitled *Mathematical Moments*, a fortnightly Australian publication *Maths by Email* for middle school students, a regular Indian magazine *At Right Angles* for school students, *Plus* magazine for senior school students in the UK, along with many other examples of various kinds. There is now a range of materials online to suit a range of backgrounds and tastes, from primary school students to the educated general public. Introducing students to this world while they are at school may help them to improve the balance in their life long after they leave school, with mathematics regarded as a source of pleasure and interest, in addition to being 'useful' for many practical purposes.

6 Conclusion

Mathematics in school is often targeted carefully at the official curriculum, recognizing the close link between school courses and further studies in mathematics, unsurprisingly. However, curricula change very slowly and need to be actualized and interpreted by mathematics teachers; preparing students for the rest of their lives in the 21st century should not be left to chance. In this chapter, we have drawn attention to four other domains of 21st century living that the mathematics curriculum needs to serve, with assistance from teachers. These include the productive worker, the careful consumer, the informed citizen and the balanced person. The productive worker needs to use their mathematics education in fresh ways related to the workplace, and probably unanticipated in school. The careful consumer needs to bring to bear their mathematical knowledge to make decisions about housing, shopping, health, recreational activities, and insurance, among other aspects of modern life. The informed citizen needs to navigate the complex world of current events, presented in various forms, many of which involve persuasion but most of which require the development and refinement of numerate behavior for optimal interpretation; indeed, tomorrow's citizens may need to be both producers and consumers of information. The balanced person is able to make use of their mathematics education, and is unafraid to engage with new materials to enrich their personal life and understand their world from a range of perspectives.

Balancing all of these 21st century needs is a difficult task, not easily achieved, but requiring conscious attention from both curriculum developers and the mathematics teachers responsible for implementing the curriculum and ensuring that it addresses the needs.

References

Australian Association of Mathematics Teachers and the Australian Industry Group (2014a). *Identifying and Supporting Quantitative Skills of 21st Century Workers: Final Report.* Canberra: Commonwealth of Australia.

Australian Association of Mathematics Teachers and the Australian Industry Group (2014b). *Identifying and Supporting Quantitative Skills of 21st Century Workers: Quantitative Skills Map.* Canberra: Commonwealth of Australia.

Australian Broadcasting Commission (2015). *Bacon, sausages, ham and other processed meats are cancer-causing, red meat probably is too: WHO.* Retrieved 27 October, 2015 from http://www.abc.net.au/news/2015-10-27/processed-meats-cause-cancer-says-un-agency/6886882

Baetz, Q., & Reese, G. (2010). *The Cereal Box Problem.* Retrieved 18 September, 2015 from https://mste.illinois.edu/reese/cereal/

Birthday Problem. (n.d.). In *Wikipedia.* Retrieved 18 September, 2015 from https://en.wikipedia.org/wiki/Birthday_problem

Blastland, M., & Spiegelhalter, D. (2013). *The Norm chronicles: Stories and numbers about danger*. London: Profile Books.

Centers for Disease Control and Prevention (2015). *Colorectal Cancer Risk by Age.* Retrieved 28 October, 2015 from http:// cdc.gov/cancer/colorectal/statistics/age.htm

Datawrapper (2015). *Datawrapper*. Retrieved 25 September, 2015 from https://datawrapper.de

Department of Statistics Singapore (2015). *SingStat Mobile App.* Retrieved 16 September, 2015 from http://www.singstat.gov.sg/services/singstat-mobile-app

Gigerenzer, G. (2002). *Reckoning With Risk.* London: Penguin.

Harcombe, Z. (2015). *World Health Organisation, meat and cancer*. Retrieved 1 September, 2015 from http://www.zoeharcombe.com/2015/10/world-health-organisation-meat-cancer/

Institute for Economics and Peace (2015). *Vision of Humanity*. Retrieved, 5 December 2015 from http://www.visionofhumanity.org

International Association for Research on Cancer (2015). *IARC Monographs evaluate the consumption of red meat and processed meat.* (Press Release No. 240). Retrieved 27 October, 2015 from http://www.iarc.fr/en/media-centre/pr/2015/pdfs/pr240_E.pdf

Jericho, G. (2015). Inequality will cast a shadow over Hockey's legacy. Retrieved 23 September, 2015 from http://www.abc.net.au/news/2015-09-23/jericho-inequality-will-cast-a-shadow-over-hockeys-legacy/6797136

Kemp, M. (2010). Developing pupils' analysis and interpretation of graphs and tables using a Five Step Framework. In B. Kaur, & J. Dindyal (Eds.) *Mathematical Appplications and Modelling: Yearbook 2010.* Association of Mathematics Educators (pp. 199-218). Singapore: World Scientific.

Kemp, M. (2012). Making connections between school mathematics and the everyday world: The example of health. In B. Kaur, & T. L. Toh (Eds.) *Reasoning, Communications and Connections in Mathematics: Yearbook 2010.* Association of Mathematics Educators (pp. 261-287). Singapore: World Scientific.

Kissane, B. (2009). Popular mathematics. In C. Hurst, M. Kemp, B. Kissane, L. Sparrow, & T. Spencer (Eds.) *Mathematics: It's Mine: Proceedings of the 22nd Biennial Conference of the Australian Association of Mathematics Teachers.* (pp 125–134) Adelaide: Australian Association of Mathematics Teachers. [Available for download at http://researchrepository.murdoch.edu.au/6242/]

Kissane, B. (2012). Numeracy: Connecting mathematics. In B. Kaur, & T. L. Toh (Eds.) *Reasoning, Communications and Connections in Mathematics: Yearbook 2010.* Association of Mathematics Educators (pp. 261-287). Singapore: World Scientific.

Kissane, B. (2015). *Reading Interesting Materials*. Retrieved 25 September, 2015 from http://wwwstaff.murdoch.edu.au/~kissane/pd/reading.htm

Merow, K. (2015). *The Wrong Door, or Why Math Gets a Bad Rap.* Retrieved 24 September, 2015 from http://www.maa.org/news/the-wrong-door-or-why-math-gets-a-bad-rap

Organisation for Economic Cooperation and Development (2013). *OECD Skills Outlook 2013: First Results from the Survey of Adult Skills,* http://dx.doi.org/10.1787/9789264204256-en

Tableau Public (2015). *Viz of the Day.* Retrieved 17 September, 2015 from https://public.tableau.com/s/gallery

Tay, E. G. (2011). Think of a number: Adapting the birthday problem for the classroom. *Australian Mathematics Teacher*, *67*(4), 22-25. Available at http://eric.ed.gov/?id=EJ956714

Understanding Uncertainty (2008) *Screening for breast cancer*. Retrieved 23 September, 2015 from http://understandinguncertainty.org/node/182

Utah State University (n.d.). *National Library of Virtual Manipulatives: Stick or Switch*. Retrieved 23 September, 2015 from http://nlvm.usu.edu

World Health Organisation (2015). *Links between processed meat and colorectal cancer*. (Media Statement on 29 October 2015). Retrieved 4 December, 2015 from http://www.who.int/mediacentre/news/statements/2015/processed-meat-cancer/en/

Chapter 5

Mathematics Subject Mastery – A Must for Developing 21st Century Skills

Berinderjeet KAUR   WONG Lai Fong   Divya BHARDWAJ

To help students in Singapore thrive in a fast-changing world, the Ministry of Education has identified competencies that have become increasingly important in the 21st Century. The outcomes of which include a confident person, a self-directed learner, an active contributor and a concerned citizen. In the US, the P21 Framework for 21st century learning has stressed that no 21st century skills implementation can be successful without developing core academic subject knowledge and understanding among all students. In this chapter, we present two key elements that are necessary for developing 21st century competencies in mathematics classrooms in Singapore. The first is the use of knowledge-building mathematical tasks and the second is teaching for metacognition. Both these elements must be used in tandem to create classroom discourse that must culminate in students actively engaging in critical thinking, problem solving, working collaboratively and articulating their thoughts and creating knowledge through their explorations.

1 Introduction

To help students in Singapore thrive in a fast-changing world, the Ministry of Education has identified competencies that have become increasingly important in the 21st Century. The outcomes of the competencies are:

- a confident person who has a strong sense of right and wrong, is adaptable and resilient, knows himself, is discerning in judgment, thinks independently and critically, and communicates effectively;
- a self-directed learner who questions, reflects, perseveres and takes responsibility for his own learning;
- an active contributor who is able to work effectively in teams, is innovative, exercises initiative, takes calculated risks and strives for excellence; and
- a concerned citizen who is rooted to Singapore, has a strong sense of civic responsibility, is informed about Singapore and the world, and takes an active part in bettering the lives of others around him. (Ministry of Education, 2010)

In the United States, the P21 Framework for 21st century learning (Partnership for 21st Century Skills, 2009) has stressed that no 21st century skills implementation can be successful without developing core academic subject knowledge and understanding among all students. The P21 framework emphasises that students who can think critically and communicate effectively must build on a base of core academic subject knowledge. Therefore core academic subjects, such as mathematics, are a bedrock component of the P21 Framework as all 21st century skills can and should be taught in the context of core academic subjects. The framework notes that 21st century skills outcomes such as critical thinkers, problem solvers, good communicators, good collaborators, information and technology literate, flexible and adaptable, innovative and creative, globally competent and financially literate individuals are a consequence of teaching, learning and assessment in core academic subjects such as mathematics.

The framework for school mathematics in Singapore has mathematical problem solving as its primary goal and the five components, i.e., concepts, skills, processes, attitudes and metacognition, guide teachers in achieving the goal (Ministry of Education, 2012). The outcomes of both the Programme for International Student Assessment (PISA) of 2009 (OECD, 2010) and 2012 (OECD, 2013) and the Trends in International Mathematics and Science Study (TIMSS) of 2011 (Mullis, Martin, Foy, & Arora, 2012; Kaur, Areepattamannil, & Boey, 2013), and 2007 (Mullis, Martin, & Foy, 2008; Kaur, Boey, Areepattamannil, & Chen, 2012) for Singapore show that the majority of Singapore students are very good in applying their knowledge in routine situations and this is definitely a consequence of what teachers do and use during their mathematics lessons. For students in Singapore to scale greater heights teachers need to nurture metacognitive learners who are active and confident in constructing mathematical knowledge. A significant finding from the CORE 2 research at NIE led by Hogan is that amongst the secondary three and primary five mathematics lessons that were studied teachers appeared to engage students in doing performative tasks (77.3% for secondary 3 and 63.7% for primary 5) more often than knowledge building tasks (22.7% for secondary 3 and 36.3% for primary 5) (Hogan, Towndrow, Chan, Kwek, & Rahim, 2013). A performative task mainly entails the use of lower order thinking skills such as recall, comprehension and application of knowledge while a knowledge building task calls for higher order thinking skills such as synthesis, evaluation and creation of knowledge.

Hattie (2009), drawing on 50,000 research articles and related achievement of 240 million students, notes that the greatest source of variance in the learning equation comes from teachers. Therefore it is important for teachers in Singapore who are desirous of improving student learning in mathematics and also developing 21st century competencies, to place emphasis on the use of knowledge building tasks and also metacognition in their mathematics lessons.

2 Mathematical Tasks

As part of an on-going professional development project, Teaching for Metacognition, that involves forty secondary mathematics teachers from seven secondary schools in Singapore, teachers were surveyed about their use of performative tasks and knowledge-building tasks in their mathematics lessons. Figure 1 shows the survey item and Table 1 shows the responses of the teachers.

<table>
<tr><td colspan="4">Mathematical Tasks
The following are examples of performative and knowledge building tasks.</td></tr>
<tr><td colspan="4">Topic: Scales and Maps</td></tr>
<tr><td colspan="2">Performative task</td><td colspan="2">Knowledge-building task</td></tr>
<tr><td colspan="2">The scale of map A is 1: 40 000
A rectangular field is 3 cm by 2 cm on the map. Find the actual area of the field in km^2.

If the area of the field is now represented on map B with scale 1: 20 000, what is the area on the map.</td><td colspan="2">The scale of map A is 1: 40 000
A rectangular field is 3 cm by 2 cm on the map. Find the actual area of the field in km^2.

If the field is now represented on map B with scale 1: 20 000, without computing any area, explain how will the size of the field be different on map B.</td></tr>
<tr><td colspan="4">Topic: Quadratic graphs and graphical solutions of simple quadratic equations</td></tr>
<tr><td colspan="2">Performative task</td><td colspan="2">Knowledge-building task</td></tr>
<tr><td colspan="2">Draw the graph of $y = x^2 - 2x - 3$ for $-2 \leq x \leq 4$.
Hence solve the equation $x^2 - 2x - 3 = -2$ graphically.</td><td colspan="2">Draw the graph of $y = x^2 - 2x - 3$ for $-2 \leq x \leq 4$.
Using your graph determine the number of solutions the equation $x^2 - 2x - 3 = a$ has.</td></tr>
<tr><td colspan="2">How often do you use performative tasks?
In ten consecutive lessons you would have used them</td><td colspan="2">How often do you use knowledge-building tasks?
In ten consecutive lessons you would have used them</td></tr>
<tr><td>Please tick the appropriate response</td><td>√</td><td>Please tick the appropriate response</td><td>√</td></tr>
<tr><td>In all the lessons</td><td></td><td>In all the lessons</td><td></td></tr>
<tr><td>In 7 - 9 of the lessons</td><td></td><td>In 7 - 9 of the lessons</td><td></td></tr>
<tr><td>In 5 - 6 of the lessons</td><td></td><td>In 5 - 6 of the lessons</td><td></td></tr>
<tr><td>In 2 - 4 of the lessons</td><td></td><td>In 2 - 4 of the lessons</td><td></td></tr>
<tr><td>In 0 - 1 of the lessons</td><td></td><td>In 0 - 1 of the lessons</td><td></td></tr>
</table>

Figure 1. Pre-intervention survey item on mathematical tasks used by teachers

Table 1

Responses of teachers to survey item on mathematical tasks

How often do you use **performative tasks**? In ten consecutive lessons you would have used them		How often do you use **knowledge-building tasks**? In ten consecutive lessons you would have used them	
	N (%)		N(%)
In all the lessons	20 (50.0)	In all the lessons	0 (0.0)
In 7 - 9 of the lessons	18 (45.0)	In 7 - 9 of the lessons	1 (2.5)
In 5 - 6 of the lessons	2 (5.0)	In 5 - 6 of the lessons	4 (10.0)
In 2 - 4 of the lessons	0 (0.0)	In 2 - 4 of the lessons	19 (47.5)
In 0 - 1 of the lessons	0 (0.0)	In 0 - 1 of the lessons	16 (40.0)

From Table 1, it is apparent that teachers were using significantly more performative tasks compared to knowledge building tasks during their lessons. This may have been a consequence of several factors, such as i) the lack of knowledge building tasks commonly found in textbooks used by the teachers; ii) inability to craft knowledge building tasks using textbook tasks that focus on direct application of knowledge and lastly iii) the push to develop procedural fluency after the introduction of concepts so as to perform routine tasks with ease during tests.

During the first two meetings of the project teachers were engaged in crafting mathematical tasks workshops. They worked in groups and crafted knowledge building tasks. Typical textbook type performative tasks were provided by the facilitators of the workshops to illuminate a focus of the content knowledge for designing the knowledge building tasks. Figures 2 and 3 show some of the knowledge building tasks crafted by the teachers.

The performative task in Figure 2 shows that the content focus is percentages. This task merely seeks the computation of a percentage and the ability to do it correctly does not provide evidence that the student has acquired any depth of knowledge about percentages. Such a task is also not sufficiently rich to engage students in any classroom discourse that would facilitate the nurturing of 21st century skills, such as critical thinking by making assumptions and presenting cases to support the argument drafted. However, tasks 1.1, 1.2 and 1.3 shown in Figure 2, provide students with opportunities to work collaboratively, engage in critical thinking and problem solving and articulate their reasoning to support their solutions based on assumptions they may have adopted. Similarly, the performative task in Figure 3 shows that the content focus is quadratic equations and the ability to solve the equation $2x^2 - 5x + 2 = 0$ does not provide us with information about the depth of students' understanding of quadratic equations. However, tasks 2.1, 2.2, 2.3 and 2.4 provide both the students and teachers with a wider scope of knowledge work that may be carried out during mathematics lessons.

<table>
<tr><td colspan="2">Performative task

30% of A is 18.
Find A.</td></tr>
<tr><td colspan="2">Knowledge building tasks</td></tr>
<tr><td>Task 1.1
Yingying makes a proposal about an exchange of monthly savings.
"I will give you 100% of what I have and all I ask for is 10% of what you have."
Will you accept Yingying's proposal?</td><td rowspan="2">Task 1.3
Two shops are having end of season sales. Their advertisements are as follows:
<table><tr><td>Shop A
Buy Two Get One Free !*

*Free item will be the cheapest among the three</td><td>Shop B
30% off all items!</td></tr></table>
Explain clearly which shop offers a greater percentage savings? State the assumptions you have made in arriving at your answer.</td></tr>
<tr><td>Task 1.2
Sally writes the following two statements.
<u>Statement 1</u>:
If A is 20% more than B, then B is 20% less than A.
<u>Statement 2</u>:
If A is 80% of B, then B is 120% of A.
Do you agree with the statements Sally wrote?</td></tr>
</table>

Figure 2. Percentage - Knowledge building tasks created by teachers

<table>
<tr><td colspan="2" align="center">Performative task

Solve $2x^2 - 5x + 2 = 0$</td></tr>
<tr><td colspan="2" align="center">Knowledge building tasks</td></tr>
<tr><td>Task 2.1
Consider $2x^2 + \mathbf{a}x + 2 = 0$
What are the possible values of a such that this equation can be solved by factorisation?</td><td>Task 2.3
Does every quadratic equation have 2 solutions? Explain your answer/s with examples.</td></tr>
<tr><td>Task 2.2
$2x^2 - 5x + \mathbf{a} = 0$ has real solutions. What are the possible values of a?</td><td>Task 2.4
If x=2 and x=3 are the solutions of a quadratic equation, what is the quadratic equation? Is there only one possible equation?</td></tr>
</table>

Figure 3. Quadratic Equations - Knowledge building tasks created by teachers

3 Teachers' Perceptions about Metacognition

The school mathematics framework in Singapore clarifies metacognition as follows:

> Metacognition, or "thinking about thinking", refers to the awareness of, and the ability to control one's thinking processes, in particular the selection and use of problem-solving strategies. It includes monitoring of one's own thinking, and self-regulation of learning. To develop metacognitive awareness and strategies, and know when and how to use the strategies, students should have opportunities to solve non-routine and open-ended problems, to discuss their solutions, to think aloud and reflect on what they are doing, and to keep track of how things are going and make changes when necessary. (Ministry of Education, 2012, p. 17)

The pre-intervention survey of the Teaching for Metacognition project sought perceptions of teachers about their understanding of metacognition and teaching for metacognition. The survey item is shown in Figure 4. All the forty teachers responded to the item. The responses are analysed and presented in the following sub-sections.

> Our mathematics syllabus states that metacognition is "thinking about thinking"?
>
> a) What does metacognition mean to you? What is your understanding of metacognition?
>
> b) Do you engage your students in metacognition during mathematics lessons? Yes / No.
> If Yes, how do you engage your students in metacognition during mathematics lessons? Give an example of what you do.

Figure 4. Survey item on teachers understanding of metacognition

3.1 *What does metacognition mean to you? What is your understanding of metacognition?*

Some samples of the teachers' responses to the first sub-item of the survey about their understanding of metacognition are as follows. To anonymise the teachers' names, codes are given as per the first alphabet of the respective teacher's school name and the teacher number, out of the total forty teachers participating in the project.

A-2 *Metacognition is an internal process in which we engage in an inner dialogue to justify our algorithms, reasons / evaluate different strategies to be used. Through metacognition, it allows one to understand deeper and also helps us reflect and improve.*

D-16 *Reflection on what is taught & learnt during lessons. Asking why I made the mistake. Being aware of what I know and what I don't know.*

N-19 *Metacognition describes the process or activities when one reflects on own's thinking, keeping track, making "deliberate" effort in organising one's thought when engaging in certain thinking activities. The individual should be able to (or at least) organise their thought process and have the ability and awareness to monitor or critique/reflect on their thinking.*

O-26 *Metacognition is higher–order thinking that enables understanding and analysis of the cognitive processes. It means setting or crafting activities to engage students in their learning.*

The qualitative responses were analysed using content analysis. The responses to each question were first scanned through for common themes, following which codes were generated and the data coded. Inevitably "a progressive process of sorting and defining and defining and sorting" (Glesne, 1999, p. 135) led to the establishment of the list of codes. The codes belong to two main themes: monitoring of one's own thinking and self-regulation of learning. Table 2, shows the list of codes and their frequencies.

Table 2
Codes and respective frequencies about what is metacognition

No.	Code	Frequency N (%)
1	**Monitoring of one's own thinking** ("thinking about thinking")	
1a	Ability to control one's own thinking processes	2 (5.0)
1b	Thinking aloud	2 (5.0)
1c	Higher order thinking / Engaging in cognitive processes during problem solving / Application of concepts taught	24 (60.0)
1d	Making connections	2 (5.0)
2	**Self-regulation of learning**	
2a	Creating awareness of thinking	9 (22.5)
2b	Guided or self-evaluation / cross examine own thinking process (what ifs)	6 (15.0)
2c	Reflect and improve / Critique own answers and thoughts	8 (20)

For the first theme, monitoring of one's own thinking, the elaborations of the four codes are as follows:

- Ability to control one's own thinking processes: This is the process of rationalising and reasoning one's thoughts, so as to critically analyse and ascertain the validity of thought (s).
- Thinking aloud (Articulation): This occurs when one expresses his or her thoughts aloud. When working with peers on complex tasks, this facilitates the sharing of ideas. When solving problems, one may ask him or herself questions such as: "What is my problem?"; "How can I do it?"; "Am I using my plan?"; and "How did I do?"
- Higher order thinking/ Engaging in cognitive processes during problem solving: This occurs during problem solving when existing (prior) knowledge is applied to an unfamiliar situation to gain new knowledge. It is the ability to analyse one's own problem-solving strategies (e.g. what strategies to be used, thinking of the steps in deriving the answer, application of concepts taught).
- Making connections: This involves making connections between mathematical concepts and between mathematical concepts and procedures; and making linkages among mathematical ideas, between mathematics and other subjects and between mathematics and everyday life.

For the second theme, self-regulation of learning, the elaborations of the three codes are as follows:

- Creating awareness of thinking: This occurs when one checks the reasonableness of an answer, seek alternative ways of solving the same problem, abandon an unproductive problem-solving strategy and adopt a more productive one; Is aware of what one knows and also what one does know.
- Guided or self-evaluation / cross-examine own thinking process (what ifs): This occurs when one uses, for example, checklists to

evaluate one's thinking processes. Over time, self-evaluation becomes a habit and will increasingly be applied more independently.

- Reflect and improve / critique own answers and thoughts: This occurs when one reflects on one's own work and improves on it. One may be guided in critiquing one's work by the following questions - "What exactly are you doing?", "Why are you doing it?" and "How does it help you?". Gradually, it becomes a matter of practice for oneself to self-question and improve his or her problem-solving skills and operation on a metacognitive level.

From Table 2, it is apparent that teachers in the project had some knowledge about metacognition but their understanding was not comprehensive. Most of them generally tended to associate metacognition with higher order thinking and problem-solving. Only a few of them associated metacognition with awareness of thinking and reflection and critiquing of one's own thoughts.

3.2 *Do you engage your students in metacognition during mathematics lessons? Yes / No. If Yes, how do you engage your students in metacognition during mathematics lessons? Give an example of what you do.*

Thirty-two of the teachers claimed that they engaged their students in metacognition during mathematics lessons. All of the thirty two gave examples of how they did so. The following are some samples of their responses.

A-3 *I will ask students to go to the board to present their answers and explain to the class. They can use this chance to articulate their thoughts and open themselves up for feedback from their peers. I will then ask him / her to clarify some points and explain to the class how is the working or answer was derived.*

N-21 *I request them to reflect on their approach to their solution. I engage them to explain why they perform a step. When they have to explain, they have to start "examining" their own thinking before they can articulate their thoughts.*

C-36 *I often pose questions that do not have a definite answer or even questions that often have no/multiple solution, e.g., (2/x) -1 = 0. Instead of asking students to think about what values of x satisfy the equation, I say "solve the equation for x". This way I do not trigger the students to think first, as it would mean I am intervening with their independent metacognition process. Only if there are students who are unable to appreciate the value of the question I would then intervene and ask a second question, "what values of x will make the LHS = 0"*

Y-43 *Asking students to explain why they feel that their solutions should be presented in the way they presented. Get them to present in an alternative way to check if they understand their own thought process.*

The qualitative responses were analysed using content analysis. Table 3 shows the codes that were arrived at and their respective frequencies. From Table 3, it is apparent that teachers engaged their students in activities that were metacognitive in nature based on their conceptions of what metacognition is as shown in Table 2. Most of the teachers engaged their students in problem solving and the use of higher order thinking skills to solve mathematical tasks.

4 Developing 21st Century Skills in Mathematics Classrooms

The forty teachers who are involved in the Teaching for Metacognition project are engaged in activities that support the development of 21st century skills and ultimately 21st century competencies in their mathematics classrooms. They are crafting knowledge-building tasks, and enacting them in their lessons such that students are engaged in metacognitive strategies. At the start of the project, as shown by the data in Tables 2 and 3, it was apparent that their knowledge of how to develop a culture of metacognition in their classrooms was limited. Therefore, the project engaged the teachers in enlarging their knowledge about how they can engage their students in metacognition. They were introduced to ten learning strategies which are adopted from Thinking about Thinking:

Table 3

Codes and respective frequencies of how teachers engage their students in metacognition

No.	Description of Codes	Frequency N (%)
1	**Monitoring of one's own thinking**	
a	Expose students to thinking skills and heuristics to solve problems / Identifying patterns / Breaking down of problems into simpler parts to solve / Apply what they have learnt / Comparing different methods / Providing counter examples / Deriving formulas / Highlight theory behind application.	27 (67.5)
b	Encourage students to think aloud the strategies and methods to solve problems.	2 (5.0)
c	Provide students with problems that require planning (before solving) and evaluation (after solving).	3 (7.5)
2	**Self-regulation of learning**	
a	Encourage students to seek alternative ways of solving the same problem.	2 (5.0)
b	Encourage students to check the appropriateness and reasonableness of the answer / justify their answers.	3 (7.5)
c	Facilitate group / whole class discussions for students to explain different methods used in solving problems.	4 (10.0)
d	Facilitate group / individual presentation for students to explain their solutions.	2 (5.0)
e	Encourage students to ask questions and clarify their thoughts during whole class instruction.	5 (12.5)
f	To reflect on their work / Engage students in peer assessment / Feedback.	6 (15.0)

Metacognition, Session 9 in The Learning Classroom: Theory into Practice (Darling-Hammond, Austin, Cheung, & Martin, 2001). The strategies and accompanying examples of activities that illuminate the learning strategies are:

1. *Predicting outcomes*

Students are asked to predict who will be the winner when a game is played before they actually play the game to investigate the outcome of the game. Students are asked to compare the outcome of the game with their initial prediction. If the outcome is different from their prediction, they will look back at their initial thoughts or possible assumptions / misconceptions made; if the outcome is similar to their prediction, they will think about the conditions / information on which they based their prediction.

2. *Evaluating work*

Students are asked to review their performance in tests or graded assignments. Students are to identify what their misconception(s)/error(s) is/are when they have not answered a question correctly and also to reflect on and determine how they can avoid making the same misconception/error in future.

3. *Questioning by the teacher*

The teacher asks students as they work "Do you understand what you are supposed to do?", "What is the information / condition given in the question that prompt you to take this step?" Teacher asks students when they give an answer – "How do you know you are right/wrong?", "Can you justify your answer?", "Is there a better or a more elegant way of obtaining the answer?"

4. *Self-assessing*

One way of self-assessing is through journaling. Students are asked to write journals based on prompts given by the teacher or simply write freely about their thoughts and feelings about their learning of (a topic in) mathematics. As they write they are able to self-assess their learning.

5. *Self-questioning*

Students ask themselves a series of questions while they work. Students can use questions to check their understanding and to help them solve a problem. They can ask themselves a series of questions such as "What is the question asking for?", "What are the conditions given in the question?", "What are the possible heuristics I may use to solve this problem?" When they ask questions while they work, students will be able to direct and clarify their thinking.

6. *Selecting strategies*

Students are asked to decide which of two or more strategies is best, for example, the method of substitution or the method of elimination for a given pair of simultaneous equations. When students decide which strategy is useful for a given task, they will have to understand the problem in order to justify the choice of the strategy.

7. *Using directed or selective thinking*

Students are required to identify a series of triangles needed to solve a trigonometric problem; or students are required to draw a roadmap of steps required to prove a geometrical relationship. This process helps students to understand the problem, identify the given information and plan the next/series of step(s) to take.

8. *Using discourse*

Students first work on a problem individually before coming together in a pair/group to compare their answers. In the pair/group, each student is to explain how he/she obtains the solution and will have to justify and convince the other(s) of the correct solution. This process helps students to concretise their thinking as they are able to hear their own thinking "visibly". It also helps students to hear others' thinking and either identify gaps in their own thoughts or learn alternative ways of explaining the same concept.

9. *Critiquing*

Students are asked to present their solution on the board and the rest of the class will provide (constructive) feedback about the work. This process allows students (who are giving feedback) to practice reading and understanding a piece of mathematical work, compare the solution presented with their own to evaluate how is one solution "better" than the other. It also allows students (who are receiving feedback) to identify the gaps in their solution and to improve their own thinking process.

10. *Revising*
Students are shown a better/more efficient/more elegant approach (as compared to their own approach) to solve a problem. After learning the alternative approach, students will then make revision to their workings. This process allows students to take note of why the alternative approach is better, to check their use of heuristics, and to identify their learning gaps.

5 Conclusion

In this chapter we have introduced the idea of knowledge-building tasks and also illustrated how teachers may craft such tasks using typical textbook questions as their starting point. We have also illustrated with examples how teachers may engage students in metacognition through ten distinct approaches that we have adopted from Darling-Hammond et al. (2001). We hope teachers will be inspired to create a metacognitive culture in their mathematics classrooms, so that their students will develop 21st century skills and have the necessary competencies for work and life in the global arena.

Acknowledgement

This chapter is based on the Teaching for Metacognition project that is funded by the Academies Fund of the Ministry of Education (AFD 02/14 BK). The authors acknowledge the contributions of the teachers in the project that forms part of the content of the chapter.

References

Darling-Hammond, L., Austin, K., Cheung, M., & Martin, D. (2001). Thinking about Thinking: Metacognition. In Darling-Hammond, LD., Austin, K., Orcutt, S., & Rosso, J. (Eds.) The Learning classroom: Theory into practice (pp 156-172). A telecourse for teacher education and professional development. Stanford University school of Education.

Glesne, C. (1999). Becoming qualitative researchers: An introduction (2nd Ed.). New York: Longman.

Hattie, J.A.C. (2009). *Visible learning: A synthesis of over 800 meta-analyses relating to achievement.* London: Routledge.

Hogan, D., Towndrow, P., Chan, M., Kwek, D., & Rahim, R.A. (2013). CRPP Core 2 research program: Core 2 interim final report. Singapore: National Institute of Education.

Kaur, B., Boey, K.L., Areepattamannil, S., & Chen Q. (2012). *Singapore's Perspective: Highlights of TIMSS 2007*. Singapore: Centre for International Comparative Studies, National Institute of Education.

Kaur, B., Areepattamannil, S., & Boey, K.L. (2013). *Singapore's Perspective: Highlights of TIMSS 2011*. Singapore: Centre for International Comparative Studies, National Institute of Education.

Mullis, I.V.S., Martin, M.O., & Foy, P. (2008). *TIMSS 2007: International Mathematics Report*. Chestnut Hill, MA: TIMSS & PIRLS International Study Centre, Boston College.

Mullis, I.V.S., Martin, M.O., Foy, P., & Arora, A. (2012). *TIMSS 2011: International Mathematics Report*. Chestnut Hill, MA: TIMSS & PIRLS International Study Centre, Boston College.

Ministry of Education, Singapore (2010). *MOE to enhance learning of 21st century competencies and strengthen art, music and physical education.* Retrieved 5 September, 2015 from www.moe.gov.sg

Ministry of Education, Singapore (2012). *O-Level, N(A) Level, N(T) level mathematics teaching and learning syllabuses*. Singapore: Author.

OECD (2010). *PISA 2009 Results: What students know and can do: Student performance in reading, mathematics and science (Volume 1)*. OECD Publishing.

OECD (2013). *PISA 2012 Results: What students know and can do: Student performance in mathematics, reading and science (Volume 1)*. OECD Publishing.

Partnership for 21st Century Skills. (2009). *Assessment: A 21st century skills implementation guide.* Tucson, AZ: Author.

Chapter 6

Teaching in the 21st Century Mathematics Classroom: Metacognitive Questioning

Cynthia SETO

The Singapore Mathematics Curriculum postulates that metacognition is one of the five key competencies for successful problem solving. Metacognitive behaviours, which involve an awareness of, monitoring of and regulating of cognitive resources during problem solving, supports the development of 21st century competencies. For students to be aware of their cognitive processes and effectively monitor and regulate these processes in learning mathematics, teachers need to provide explicit guidance and model these processes in their classrooms. This chapter focuses on the use of questions to provide opportunities for students to think aloud through an articulation of their problem-solving processes, thus making their thinking visible and creating a greater level of awareness of their cognitive processes. This helps students to better monitor their own cognitive activities during problem solving, and to regulate their problem solving processes. It aims to provide teachers with greater understanding of metacognitive behaviours and build teachers' confidence to develop students to be metacognitive learners.

1 Introduction

To prepare our students to be future-ready, the Ministry of Education has identified competencies that are increasingly important in the 21st Century. The development of 21st Century Competencies will help our students to embody the Desired Outcomes of Education (Ministry of Education, 2010). One of these outcomes of education is to develop each student to be a self-directed learner who questions, reflects, perseveres and takes responsibility for his/her own learning. These attributes are aligned to the description of metacognition in the Singapore Mathematics Curriculum Framework as shown on Figure 1.

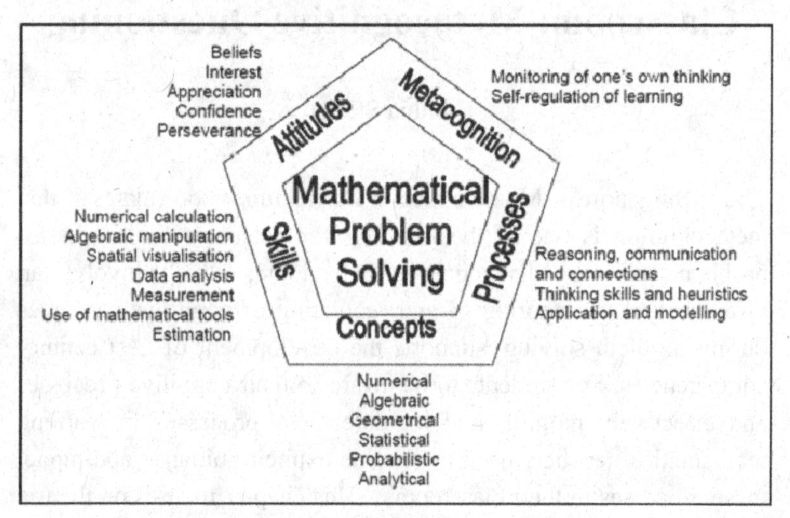

Figure 1. The Singapore Mathematics Curriculum Framework

The framework specifically states that metacognition is basically "thinking about thinking" (Ministry of Education, 2007, p. 12). It refers to the awareness of, and the ability to control one's thinking processes, in particular the selection and use of problem-solving strategies. This includes monitoring of one's one thinking, and self-regulation of learning. The development of mathematical problem solving ability requires a good understanding of mathematical concepts, proficiency in mathematical skills and processes, a positive attitude towards mathematics and an awareness of one's thinking processes.

The processes refer to the mental processes involved in comprehending and gaining mathematical concepts. These cognitive processes include thinking, reasoning, communicating and applying mathematical ideas. Metacognition is the knowledge and understanding of our cognitive processes and abilities and those of others. It also includes regulation of these processes. In other words, metacognition is the knowledge we hold about our own thinking, and the thinking of other people.

2 Metacognition in Problem Solving

Metacognition plays an essential role in mathematical problem solving and it has been recognised as one of the most relevant predictors of accomplishing complex learning tasks (Dignath & Buttner, 2008; Van der Stel & Veenman, 2010). Developing students' abilities in mathematical problem solving is the central focus of the Singapore mathematics curriculum and metacognition is one of the key factors in facilitating success in problem solving. Hence, it is important that teachers empower their students to be aware of their own problem solving process, take control of this process, and seek help whenever necessary.

How do teachers promote such metacognitive awareness and behaviours in their classrooms? Lee (2009) highlighted four instructional strategies that have been found to be effective in helping students to be more aware of their thinking process. They are mathematical log writing, effective questioning techniques, identification of structural properties of problem, and pair/group problem solving. He also conceptualised the Problem Wheel (Lee, 2008, p. 65) as a basis for questioning. The interactivity of the various components as shown in Figure 2 reflects the dynamics involved in solving word problems. It also serves as a bridge for students to make an effective connection between the understood information and the mathematical knowledge they have acquired in order to translate information in a word problem into mathematical concept.

Based on Polya's four-stage model (Polya, 1957) in problem solving, the Problem Wheel facilitates students to understand the word problem and in coming out with a plan to find the solution to the problem. The four stages in the Polya's Model are: *understanding the problem* (identifying the givens, wanted and conditions), *devising a plan* (making connections to existing knowledge), *carrying out the plan* (checking each step), and *looking back* (checking the solution and looking for alternative methods). The Problem Wheel seeks to make the processes of 'understanding' and 'devising a plan' even more explicit through the use of picture or diagram to represent the problem as well as identifying the topic and formula. As such, the Problem Wheel guides teachers to focus students' attention to better understand and make sense of a given problem (Lee, Yeo, & Hong, 2014). The purpose and examples of question prompts for each of the component in the Problem Wheel are presented in Table 1.

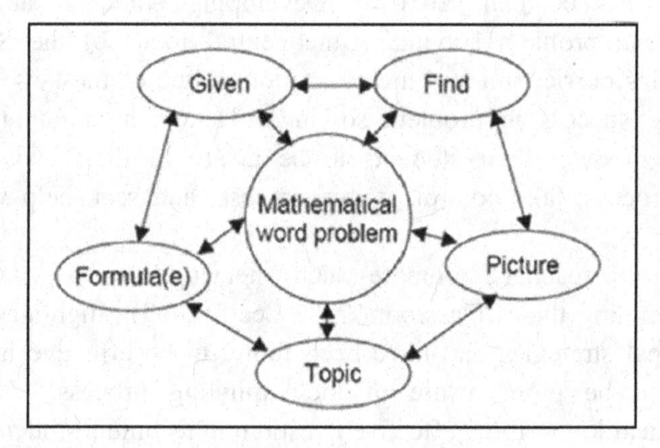

Figure 2. The Problem Wheel

When teaching students to solve word problems, teachers can use the question prompts in Table 1 to think aloud so as to model the problem solving process for students. When teachers verbalise their thoughts during problem solving, students can see how the mind responds to thinking through difficult information and constructing meaning from the problem situation. In this way, teachers demonstrate practical ways of approaching a word problem and make the thinking processes that underlie mathematical problem solving visible to students. By verbalising their inner speech as they think through a problem, teachers model how expert thinkers solve a problem.

Table 1

Purpose and examples of question prompts in the Problem Wheel

Component	Purpose	Examples of Question Prompts
Given	Raise awareness of given information provided within the context of the problem	What is the problem about? What information is given in the problem?
Find	Encourage self-directed question on the information that needs to be found	What are you supposed to find? What unknown information can I find?
Picture	Serve as a self-regulatory function to develop relationship between the known(s) and the unknown(s)	What pictures or diagram can you draw to represent the problem? How do you draw the diagram?
Topic	Act as a bridge between the problem and the content learnt	Which topics are relevant to this problem?
Formula (e)	Act as a bridge between the problem and students' ability to select appropriate mathematical skills and knowledge to solve the problem	Which is the most appropriate formula(e) to solve this problem? What equation can I write to link the known and unknown?

Teachers need to provide students with opportunities to practise the technique of thinking aloud, either in pairs, small groups or individually. Thinking aloud with teachers and with one another encourages students to gradually internalise this dialogue which is the means by which they direct their own behaviours and problem solving process (Tinzmann et al., 1990). Students with metacognitive skills have the ability to use prior knowledge to plan a strategy for approaching a word problem, use the appropriate strategies to problem solve, reflect on and check their solutions, and self-correct their approach as needed. Flavell (1976), who coined the term 'metacognition', offers the following example:

> I am engaging in metacognition if I notice that I am having more trouble learning A than B; if it strikes me that I should double check C before accepting it as fact. (p. 232)

3 Metacognitive Questioning in the Classroom

Metacognitive strategies can be taught to students to improve their learning (Dignath & Buttner, 2008). This is further supported by Schoenfeld (1992) who found that successful mathematics learners are metacognitively active. According to Mevarech and Kramarski (2014), modelling through thinking aloud is one of the best ways to make students aware of the strategies. They also highlighted the need to ask self-directed questions which serve two purposes: first, it encourages students to articulate their problem solving strategies, and second, it fosters reflections on these activities. Empowering students to spontaneously ask themselves questions might lead them to think about their thinking, and regulate and monitor their own cognitive processes. And, a teacher needs to show students how to ask self-directed questions and provide students with opportunities to practice them. This chapter presents two vignettes to illustrate how teachers can model and provide explicit guidance for students to embody metacognitive questioning in their mathematics classrooms.

The first vignette narrates a Primary 5 Mathematics lesson for a class of low progress learners. It describes a teacher, Mr Tan, who used the think aloud strategy to model the use of questions to solve the word problem below.

> Alice needs $\frac{1}{3}m$ of ribbon to tie a present. Vivian needs $\frac{4}{9}m$ of ribbon to tie her present. How much ribbon do they need altogether?

After reading the question once, Mr Tan directed his students' attention to the given information in the word problem. To do this, he used the think aloud strategy to ask himself the following questions and answered them accordingly:

- How many people are there? (There are 2 people.)
- What are their names? (They are Alice and Vivian.)
- What are they are going to do? (Each of them is going to tie a present with ribbon.)

- How much ribbon does Alice need? (Alice needs $\frac{1}{3}$ *m*.)
- How much ribbon does Vivian need? (Vivian needs $\frac{4}{9}$ *m*.)

Having identified the given information from the word problem, Mr Tan then asked himself what he was supposed to find. He pointed to "How much ribbon do they need altogether?" and read the question to himself. Following that, he self-questioned the meaning of "altogether" and articulated that it referred to the total length of the ribbon needed by Alice and Vivian. As such, it was the 'whole' that he needed to find and the parts are the length of ribbon needed by Alice and Vivian. At this point, he also highlighted the $\frac{1}{3}$ *m* of ribbon and $\frac{4}{9}$ *m* of ribbon. A model was drawn to represent this information and an addition equation was also formulated.

After modelling metacognitive questioning, Mr Tan arranged his students in pairs to practice how they could ask themselves questions to understand the problem and devise a plan to solve the problem. A set of questions based on the components in the Problem Wheel were given to students to scaffold the practice of self-questioning to solve this problem.

Winston and Derek drank $\frac{3}{4}$ *l* of milk altogether. Winston drank $\frac{1}{3}$ *l* of milk. How much milk did Derek drink?

The excerpt below shows that metacognitive questioning by the students A and B helped them make sense of the part-whole concept in the word problem to develop a plan.

A *...they got two people. Who are they? They are Winston and Derek*

B *...so, ok, good. They drank $\frac{3}{4}l$ of milk altogether. Together is what? The meaning is about the total amount. So $\frac{3}{4}l$ is ...*
So this one (pointing to Winston) drank $\frac{1}{3}$ l. So ...

A *So, ask to find how much did Derek drink?*

B *Basically, either this (pointing to $\frac{3}{4}l$) plus this (pointing to $\frac{1}{3}l$) or this (pointing to $\frac{1}{3}l$) plus this (pointing to $\frac{3}{4}l$).*

A *Draw model...so Derek, you don't know...put question mark. And then put $\frac{3}{4}$ ...*

B *Cannot be here....$\frac{3}{4}$ must be here (show on Diagram)...for the 2 people. Yes, correct. How do you do?*

A *So, $\frac{1}{3}$ plus $\frac{3}{4}$*

B *So, cannot be, can, meh....this (pointing to the model for Winston) and this (pointing to the model for Derek) is $\frac{3}{4}$. What is $\frac{3}{4}$? They drink together.*

A *Subtract ... $\frac{3}{4}$ subtract $\frac{1}{3}$.*

In translating their understanding into a model drawing as shown in Figure 3, they monitored their understanding through questioning and articulating their thoughts. It is going through this process which led them to refine their thinking to finally formulate a subtraction equation to solve the problem.

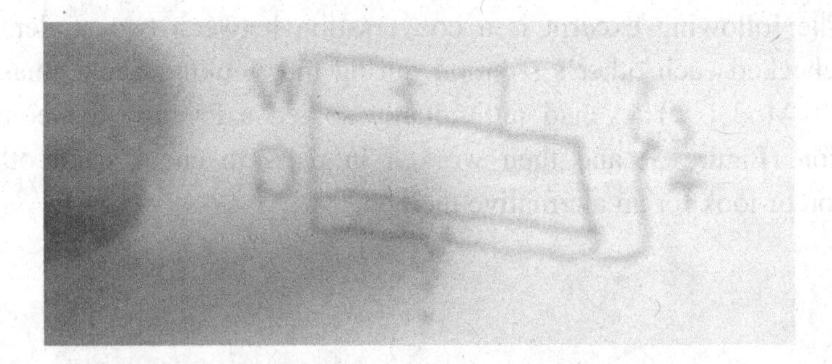

Figure 3. Model drawn by student A

In the second vignette, Miss Lim provided sentence starters during partner talk to empower students to use questions and express their thought processes more effectively. Like Mr Tan in the first vignette, Miss Lim also used the think aloud strategy to explicate the four stages in Polya's Model for solving mathematics problems. In her lessons, her students practiced self-questioning either in small-groups or individually during problem solving. While her students are proficient in asking themselves questions during problem solving, Miss Lim noticed that they had the tendency to focus only on the computation or simply browsing through their solutions during the 'looking back' stage of Polya's Model. Therefore, she included an activity for her students to verbalise their thinking processes as they checked their solutions. And, to facilitate this process, Miss Lim displayed sentence starters on the wall as shown in Figure 4.

Figure 4. Wall display of sentence starters

The following excerpt is a conversation between two students as they checked each other's solution during the 'looking back' phase of Polya's Model. They had individually solved a Primary 5 Geometry question (Figure 5) and then worked in pairs to check each other's solution or look for an alternative method.

Figure 5. Student Q pointing to the angles on a straight line

The transcript below shows students using the sentence starters and the components in Problem Wheel as they revisited their solutions. They were puzzled as to why they had different answers although their computations were correct.

P *Why different answers? Cannot be. Calculation correct. Yours ...one more step... you find angle x first. Then 180° subtract 55° and subtract 60° and got 65°.*

Q *What is given in the problem? 60° and 125°. What we need to find? Find angle c. You take 180° to minus 125° to find c. You got 55°. So you are saying that 125 plus angle c equal 180.*

P *Because, they (inaudible)....angles on a straight line. So you are saying that you cannot take 180° to minus 125° to find c.*

Q *180 is always on a straight line. You have 60° and you have 125°. You want to find angle c. Must write angle x here and find x.*

P *No, no. I disagree because angle c plus 125 are... are on a straight line.*

Q *I disagree...not c ...because you need another angle...angle x.*

P *Hmmm....They must be side by side...you mean...then counted.*

Q *You see the textbook (pointing to a page on the math textbook). Angles on a straight line. The angles must be next to it...have to be together ...like this (shown in Figure 5) ...then the sum of angles on a straight line is 180°.*

This excerpt illustrates that with explicit guidance through the Problem Wheel and the sentence starters, students are able to regulate their own cognitive processes; student P self-corrected her misconceptions while student Q reaffirmed her understanding about the sum of angles on a straight line. In providing students with opportunities to articulate their problem solving strategies to each other, Miss Lim aimed to equip every student to be metacognitive in checking their own solutions.

4 Conclusion

Questions facilitate students to think aloud to reflect and be aware of one's own knowledge – what one does and does not know – and one's ability to understand, control, and manipulate one's thinking processes. This includes knowing where and how to use particular strategies for problem solving. To promote metacognition in our mathematics classrooms, teachers need to provide tasks at an appropriate level of difficulty. This means the task has to be challenging enough so that students need to apply metacognitive strategies, such as self-questioning, to monitor success but not so challenging that students become overwhelmed or frustrated. Teachers need to prompt students to think about what they are doing during problem solving.

Improving metacognitive skills is not a "magic bullet" to solve problems faced by our diverse range of learners in our classrooms. Nor is it a quick fix. Helping our students to become more metacognitively aware is a deliberate process. However, it is an essential approach to learning how to learn. It is also one of the toolkits of strategies that will be of relevance not only in school but also in preparing for a productive life in the 21st century.

Acknowledgement

The author is grateful to the teachers and there students, whose lessons are illustrated in this chapter, for their contribution.

References

Dignath, C., & Buttner, G. (2008). Components of fostering self-regulated learning among students. A meta-analysis on intervention studies at primary and secondary school level. *Metacognition and Learning, 3*, 231–264.

Flavell, J. H. (1976). Metacognitive aspects of problem solving. In L. B. Resnick (Ed.), *The nature of intelligence* (pp. 231–236). Hillsdale, NJ: Lawrence Erlbaum Associates.

Lee, N. H. (2008). *Enhancing mathematical learning and achievement of secondary one normal (academic) studies using metacognitive strategies.* Unpublished doctoral dissertation, National Institute of Education, Nanyang Technological University, Singapore.

Lee, N. H. (2009). *Teaching this thing called "metacognition"*. Retrieved 17 May, 2015 from http://singteach.nie.edu.sg/issue20-mathed/

Lee, N. H., Yeo, D. J. S., & Hong, S. E. (2014). A metacognitive-based instruction for primay four students to approach non-routine mathematical word problems. *ZDM – The International Journal of Mathematics Education, 46*, 465–480.

Mevarech, Z., & Kramarski, B. (2014). *Critical maths for innovative societies: The role of metacognitive pedagogies*. OECD Publishing.

Ministry of Education (2007). *Primary mathematics teaching and learning syllabus.* Singapore: Author.

Ministry of Education, Singapore (2010). *MOE to enhance learning of 21st century competencies and strengthen art, music and physical education.* Retrieved 31 December, 2015 from www.moe.gov.sg

Polya, G. (1957). *How to solve it* (2nd ed.). New York: Doubleday & Company.

Schoenfeld, A. H. (1992). Learning to think mathematically: Problem solving, metacognition, and sense-making in mathematics. In D. A. Grouws (ed.), *Handbook of research on mathematics teaching* (pp. 334–370). New York: MacMillan Publishing.

Tinzmann, M. B., Jones, B. F., Fennimore, T. F., Bakker, J., Fine, C., & Pierce, J. (1990). *What is the collaborative classroom?* NCREL: Oak Brook.

Van der Stel, M., & Veenman, M. V. J. (2010). Development of metacognitive skilfulness: a longitudinal study. *Learning and Individual Differences, 20*, 220–224.

Chapter 7

Listening and Responding to Children's Reflective Thinking: Two Case Studies on the Use of the National Assessment in Japan

Keiko HINO

In the mathematics classroom, teachers play a crucial role in deepening children's understanding and promoting thinking. As an important part of 21st century competencies, this chapter focuses on the development of reflective thinking in children, reports on two teacher education case studies, and discusses methods for improving teachers' abilities to listen and respond to children's often underdeveloped reflective thought. The case studies report on how national assessment results and resources were used to provide teachers with the opportunity to anticipate, compare, and investigate children's reflective thinking, and to guide lesson design to assist children to hone their reflective thinking skills. These case studies highlight the importance of raising teacher awareness and understanding the use of children's reflective thinking in the mathematics lesson.

1 Introduction

To prepare children for their future and work in a rapidly changing society, 21st century competencies have been identified and widely discussed. The Organization for Economic Co-operation and Development (OECD) (2005) defined these *key competencies* as the psychosocial prerequisites for a successful life and a properly functioning society in the emerging technologically driven, diverse and globalized world. The three broad key competency categories were identified as; "using tools interactively," "interacting in heterogeneous groups," and "acting autonomously." Underlying this framework is the requirement to promote reflective thought and action (called *reflectiveness*), which requires the subject to become the object of its own thought processes and implies metacognitive skills (thinking about thinking) and creative abilities, and taking a critical stance (OECD, 2005, p. 9). In the Assessment and Teaching of Twenty-First Century Skills Project, *21st century skills* were defined (Griffin, McGaw, & Care, 2012) to include problem-solving/decision-making, creative thinking, critical thinking, learning how to learn, communication/collaboration, information and ICT literacy; all of these are basic skills for living in this globalized society.

Many countries are now defining their essential future competencies, developing complementary objectives, and implementing 21st century educational policies. In Japan, the current Mathematics Course of Study emphasizes the development of mathematical activities for all school levels to foster basic knowledge and skills (through spiral sequencing), and to promote thinking/decision-making/communication abilities so as to motivate children's interest in learning mathematics (see MEXT, 2008, pp. 1–5). In 2013, the National Institute for Educational Policy Research (NIER) proposed a framework of *21st century skills* (tentatively named), which consists of three classes; *thinking abilities* at the center, *basic literacy* as the scaffold, and *practice skills* as the director (NIER, 2013). One of the major issues in the curriculum revision that is currently underway is how these 21st century competency aspects can be incorporated into the Course of Study for each subject (MEXT, 2014).

To foster mathematics competencies or skills, it is assumed in this chapter that teachers play the most crucial role. Teachers need to first understand these competencies by investigating the essential inherent values, and then designing mathematics lessons that promote their use. However, this is not an easy task, as teachers are most often concerned with teaching mathematics as knowledge rather than promoting thinking abilities through the teaching of concepts. This chapter focuses on reflective thinking as an important part of 21st century competencies or skills, and through an examination of two teacher education case studies, discusses the ways of stimulating teacher abilities to listen and respond to children's often underdeveloped reflective thinking.

2 Perspectives

2.1 *Reflective thinking and the teacher's mathematical knowledge*

Reflective thinking has been recognized as crucial for successful mathematical problem-solving and a deeper understanding of the content. "Reflection is a particular way of thinking and cannot be equated with mere haphazard 'mulling' over something" (Rodgers, 2002, p. 849, cited by Kaur, 2013, p. 4). It is metacognitive in nature and involves an awareness of, and the ability to control one's own thinking. In his famous book *How to solve it*, G. Polya proposed four problem-solving phases. The fourth phase, in particular is, *looking back*, and is related to the act of reflection as it involves a reconsideration and reexamination of the results and the path that led to those results, thus requiring the problem solvers to question themselves by answering reflective questions such as; "Can you check the result?" "Can you check the argument?" "Can you derive the result differently or see it at a glance?" "Can you use the result, or the method, for some other problem?" (Polya, 1985, pp. 14–16). These questions imply that to think reflectively, it is necessary to critically examine the obtained solution, draw on self-knowledge for further inquiry, and construct arguments, alternatives, or suggestions.

In a review of the knowledge needed to teach mathematics, Beswick (2013) stated that teachers needed all aspects of teacher knowledge to nurture reflective thinking in learners. From an analysis of three teaching and learning episodes, Beswick summarized those aspects of teacher knowledge and beliefs that were particularly relevant (pp. 74–76). Teachers need to have the following knowledge and belief related tenets:

- the mathematical knowledge to choose appropriate, challenging yet accessible tasks and to convert them into effective lessons;

- general pedagogical knowledge and pedagogical content knowledge particular to mathematics to enable the provision of suitable materials that inform the teachers of in-the-moment decisions on interactions and promote teacher-student discourse to support the students' thinking; and
- a belief that mathematics is about sense making, so as to stimulate interest in their students' thinking and beliefs about their capability to learn mathematics, as well as the teachers' own beliefs in the importance of their role as a facilitator of knowledge and thinking skills.

In this chapter, special attention is paid to the development of the teachers' abilities to listen to and interpret children's thinking and to respond to the children to ensure deeper, more complex understanding (c.f., Davis, 1997; Crespo, 2000). In terms of teacher's content knowledge for teaching (Ball, Thames, & Phelps, 2008), these abilities relate to the teacher's *knowledge of content and students* and *knowledge of content and teaching*. They can also relate to *specialized content knowledge* because "evaluating the plausibility of student claims" is regarded as one everyday task of teaching that requires that knowledge. In the Japanese approach to developing expertize in mathematics teaching, there is a consistent focus on the development of the children's thinking abilities. Takahashi (2011) argued that anticipating children's responses is especially essential for planning *neriage* (extensive discussion); "Because the discussion will change based on how students solve the problem, anticipating all the solution methods, including possible misunderstandings, helps teachers prepare to handle the discussion flexibly" (p. 212). To develop the teachers' abilities to listen and respond to children's reflective thinking, this chapter reports on two case studies in which the results of a national assessment are used.

2.2 *National Assessment of Academic Ability*

In Japan, the Ministry of Education, Culture, Sports, Science and Technology (MEXT) has conducted the *National Assessment of Academic Ability* (NAAA) since 2007 on children in the sixth year of primary school and the third year of lower secondary school. The NAAA is conducted in concert with the implementation of the current Course of Study and was aimed at the realization of the Plan-Do-Check-Action (PDCA) cycle. Mathematics is one of school subjects chosen to be assessed. The assessment is carried out by using two problem Sets. Problem Set A concerns "knowledge," and assesses the basic academic knowledge and skills needed for learning in later grade levels and knowledge indispensable for everyday life. Problem Set B concerns "application" and assesses the ability to utilize knowledge and skills to solve various problems that include everyday life problems. Children are assessed on their ability to develop plans to solve problems, to carry out these plans, and to evaluate and improve the plans. Thus, the "application" ability relates to the 21st century competencies. Specifically, Problem Set B for mathematics targets the following four abilities (NIER, 2015a):

- Observe the object and accurately understand it by paying attention to number, quantity, or figure.
- Classify and arrange the given information and/or choose relevant information properly.
- Think logically and/or reflectively.
- Interpret the phenomenon mathematically and/or express one's own thinking mathematically.

One of the features of the NAAA is the ways of analyzing and reporting the results. In addition to the basic statistical data such as mean percentage of correct answers, the children's responses to the mathematics problems are analyzed using anticipated types of response. This analysis is conducted against each problem item. Table 1 shows the anticipated types of responses together with distribution of children's responses to a problem (Figure 1).

Another feature is that on the basis of an examination of the tendencies in children's responses, both their difficulties and the areas that require teachers' attention are highlighted. Regarding the problem in Figure 1, the analyses are followed by suggestions to teachers as to the teaching of these numerical relationships in the calculation of decimal numbers. They include activities through the use of diagrams or number line representation, and learning to transpose ● with a simple number so that the children are able to construct clues by themselves to approach these types of problems (NIER, 2008, pp. 188–189). Further, to distribute the results and feedback to teachers, the NIER publishes, both in print and online, examples of effective trials at schools that have used the NAAA results to improve teaching practice as well as examples of mathematics lesson ideas using the NAAA problem items (e.g., NIER, 2015b).

Problem Set A [3], 2008 (translated by the author)

In the mathematical expressions listed below, ● expresses the same number, and is not 0.
Choose the mathematical expressions from 1 to 4 below which are larger than the number that ● expresses.

1 ● × 1.2

2 ● × 0.7

3 ● ÷ 1.3

4 ● ÷ 0.8

Figure 1. An example of problems in Problem Set A

Table 1
Ddistribution of actual responses for Item A [3], 2008

Item number		Type of response	Percentage of response (%)	Correct answer
[3]	1	Choose 1 and 4	45.3	◎
	2	Choose 1	2.9	
	3	Choose 4	1.8	
	4	Choose 1 and 2	4.4	
	5	Choose 1 and 3	12.0	
	9	Other responses	23.9	
	0	Non-response	9.7	

3 Using the NAAA with Prospective Teachers

The first case study focused on 45 second-year students at Utsunomiya University. Most students were non-mathematics majors who had not yet experienced student teaching (student teaching is allocated in the third and fourth years in the curriculum). At the last meeting of elementary mathematics methods course in July 2015, taught by the author, these students were given a NAAA problem from Problem Set B. In this section, the prospective teachers' listening and responding to children's real thinking and their learning from the activities are described.

3.1 *Prospective teachers' anticipation of children's responses*

First, the 45 students were asked to solve all three items for the Vaulting Horse problem. Figure 2 shows problem item (2). Problem items (1) and (3) are multiple-choice items, and almost all students successfully chose the correct answers. However, for item (2), 32 correctly selected 2 and wrote a satisfactory reason (71%), and four students selected 1.

Then, the students anticipated the percentage of correct answers for each item. Here again, there was a notable discrepancy in (2) between the averages for their anticipated percentages (42%), and the actual percentages for the correct answers (27%) reported in NIER (2012). They were surprised to learn a much lower percentage of correct answers with respect to this item.

The students further anticipated the children's mistakes and unsatisfactory responses to item (2). The main anticipated responses included "addition miscalculation", "writing the reason only using words (or mathematical expressions)" and "no response". Twenty-one students (47%) also anticipated that children would ignore the vaulting horse layers and choose lengths so that they would add up to 70. Indeed, this was one of the typical mistakes the children actually made (NIER, 2012, pp. 236–237). However, there was almost no anticipation of unsatisfactory responses such as, "we cannot reach 70 cm by adding layers to the medium vaulting horse," which was also a typical response given by the children.

At Yukari's school, there are two types of vaulting horse, small and medium. The top layer of the small vaulting horse is 30 cm high, and each layer second through the eighth is 10 cm in height. For the medium vaulting horse, the top layer is 35 cm high and each layer second through fourth is 15 cm in height. For each layer fifth through eighth, the height is 10 cm.

Small Vaulting Horse (8 layers) **Medium Vaulting Horse (8 layers)**

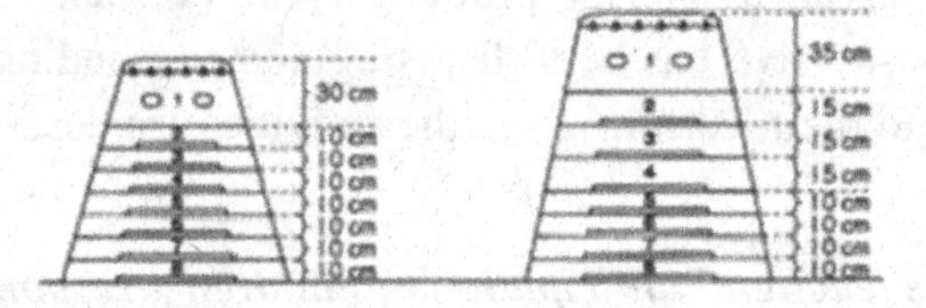

(2) Yukari's teacher asked her and her friends to make the small and medium vaulting horses the same height. First they put 5 layers on the small vaulting horse they often used for practice. The height was 70 cm.

Small Vaulting Horse (5 layers)

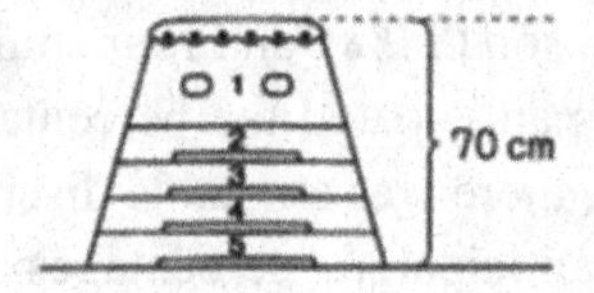

Next, they wanted to set up the medium vaulting horse so that the height was also 70 cm. Is it possible to set up the medium vaulting horse at a height of 70 cm? Select the correct answer from 1 or 2 below and write the number. Also, explain why you chose that answer using words and numbers.

1 It is possible to set up the medium vaulting horse at a height of 70 cm.

2 It is not possible to set up the medium vaulting horse at a height of 70 cm.

Figure 2. Vaulting Horse, Problem Set B 2 (2), 2012 (translated by Project IMPULS (IMPULS, 2015))

3.2 *Prospective teachers' lesson ideas using this problem item*

The students were then asked to think about how to teach a lesson using item (2) to foster the children's ability to make decisions using relevant information and to express their reasoning. At first, they thought about important teaching points by themselves, after which they formed small groups and shared their ideas. Then, the groups presented their ideas using magnetic boards. Finally, an example of the lesson idea for item (2) (NIER, 2012, pp. 238–239) was distributed and explained.

In this part of the activity, the most impressive observation was the students' attention to the use of concrete materials in the lesson. When the students individually thought about the important teaching points, 30 students wrote about the use of concrete materials, such as, "prepare an actual vaulting horse", "prepare a model of a vaulting horse" or "go to the gym and look at a vaulting horse". Twenty-four students wrote only about the use of concrete materials. Thirteen wrote about the ways of developing the lesson, but their attention was more on the teachers' explanations than on the children's thinking. During the presentation of ideas, the content written on the magnetic boards also showed a strong attention to the use of concrete materials. One group drew a model of a vaulting horse made of paper that could be manipulated during the lesson.

The example of an idea for the lesson prepared by NIER had two tasks, as shown below, and included selected interactions between the teacher and the children and an example of blackboard writing (NIER, 2012, pp. 238–239).

- Task 1: Can we make the medium vaulting horse 70 cm high?
- Task 2: Let's write down in your notebook the reason we cannot make the height of the medium vaulting horse 70 cm.

In Task 1, the selected interaction between the teacher and the children began with one child saying, "We can make 70 cm using the first, second, fifth, and sixth layers." The interaction continued using a picture of the vaulting horse and ended with the conclusion that they could not achieve the 70 cm as the fifth layer could not be put on the second layer because of the gap between them. This task can be seen to be consistent with the students' attention to the use of concrete materials.

Importantly, besides Task 1, the example included Task 2, which asked the children to think about the reason they were unable to achieve the height of 70 cm. It also included examples of the children's reasons. Selected interactions demonstrated the teacher's key questions; as in the following:

> *Let's look into your friend's reason in their notebook and discuss "which part is easy to understand" and "which part does not look like enough for the explanation".*
>
> *We do not need to write everything. Then let's think together about an explanation that is easy to understand by writing only the necessary information.*

The example of blackboard writing showed memos on "what we should write in order to explain what we cannot" and an explanation that the class constructed together.

3.3 *Their reflections*

Finally, the students wrote their reflections on the teaching of the lesson using item (2). Seventeen students again wrote about the use of the concrete materials (11 students wrote only about the concrete materials), but a greater number of students (27) wrote about ways to develop the lesson. Many of these reflections included words such as "thinking", "discussing" or "sharing". There were five reflections that explicitly stated their lack of attention to the variety of unsatisfactory (or vague) reasons and ways of building on such variety in the children's thinking (see below):

> *I tended to try to narrow the gap and lead the children to the correct response. But I thought it is also important to show children the inadequate explanations and ask them how to change them so that everyone can understand the message.*

I can easily think of going to the gym or making a model in the introductory part of the lesson. To me it is easier to conceive how to raise the children's interest. But in the main part of the lesson or in the concluding part of the lesson, I am not familiar with what I should do in order for the children to absorb the ability that I want to foster.

3.4 *Summary*

Problem item (2) requires that the decision as to why it is not possible to set the medium vaulting horse at a height of 70 cm be explained logically in a written form. Writing an explanation of an impossible result is often unfamiliar to children. This type of exercise demands reflective thinking and the children need to examine their own reasoning and construct an argument by checking the adequacy of their thinking.

Prospective teachers found that the use of the NAAA results and resources was effective in developing their abilities to carefully and critically listen and respond to the children's thinking. However, they tended to think that the explanation writing was not so difficult, and was therefore surprised at the much lower percentage of satisfactory explanations. Further, the children's typical responses, which showed their underdeveloped reflective thinking skills, made the teachers listen more critically to the way the children were seeking to solve the problems and the reasons for their consequent difficulties. The prospective teachers also tended to pay more attention to the concrete materials so as to avoid any misunderstandings when responding to the children's thoughts. Yet, while using the concrete materials as a basis can be effective, it does not allow for the children's reasoning to be highlighted as the focus of the lesson. Using an example of NIER lesson ideas, the prospective teachers were able to understand why and how the children's reasoning should be included as a main part of the lesson.

4 Using the NAAA with Practicing Teachers

The second case study focused on three teachers who have 15 to 20 years of teaching experience. Mr Takahashi is a lower secondary school teacher while Ms Hirose and Mr Nagashima are primary school teachers. The teachers participated in the six-month-long professional development program at Utsunomiya University from October 2013 to March 2014. During the program, the teachers intensively used NAAA problem items to explore their research theme of fostering the children's ability to utilize knowledge and skills to solve problems. All the teachers' activities were collaborative. The author was also involved in the activities, mainly through regular meetings. In this section, the teachers' explorations using the NAAA problem items were described.

4.1 *Teachers' assessment of their children using the Park problem*

The teachers began by collecting data on their children's current abilities to utilize knowledge and skills using NAAA problem items and additional questions. One of the problems was the Park problem (NIER, 2007, p. 171). When a path to the shop was given, item (1) asked children to draw a path of equal distance going back to Hiroshi's house. Item (2) asked them to choose the longer of the two paths given. Item (3) asked them to compare the areas of the two parks (see Figure 3).

The percentages of correct answers are shown in Table 2. The teachers obtained further information on their children's utilization of the knowledge on how to calculate the area of a parallelogram from the result of the additional questions with respect to the item (3) (Table 3).

The teachers found that the percentage of correct answers for item (3) was much lower than for the other items (Table 2). This was because many of their children (54.2%) made mistakes by choosing the slanted side (instead of the height) to calculate the area of the Central Park (parallelogram) (Table 3). This result was consistent with the one reported (NIER, 2007, pp. 175–176). Furthermore, 62.5% of the children were found to be able to write the formula for the area of the parallelogram. It was also found that of the 54.2% of children who chose the slanted side, 30.8% did not write any reason on their worksheet ("no response"). From these observations, the teachers summarized the children's difficulties as follows:

- To understand the relationship between lines (parallel and perpendicular) and the height of the parallelogram.
- To understand the formula for the area of a parallelogram.
- To accurately understand the required information from the problem statement and the map.
- To choose the necessary information from the excess information shown on the map.
- To connect the information they chose from the map with their learned knowledge.

(3) There is a park, named East Park, near Hiroshi's house. Which is larger, the area of Central Park or the area of East Park? Write your answer. Also explain the reason using words, mathematical expressions, and so on.

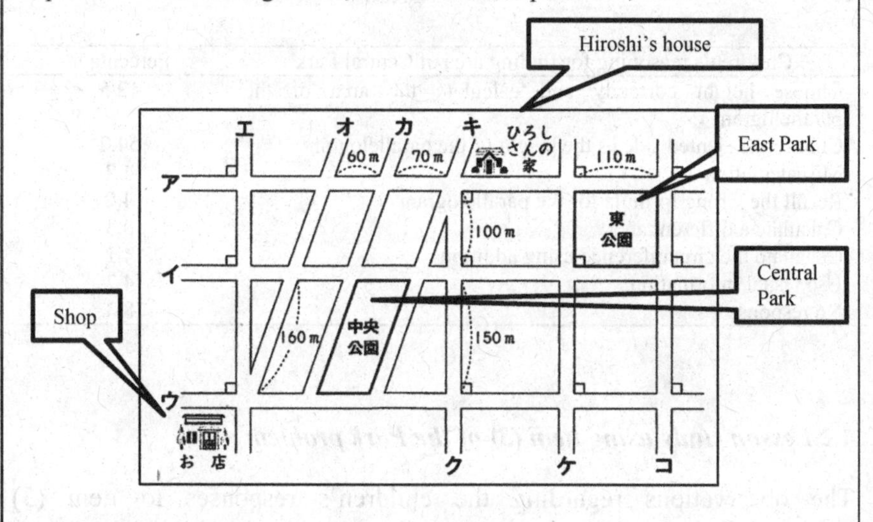

○ Roads ア，イ，ウ are parallel.
○ Roads オ，カ，キ are parallel.
○ Each of the Roads ア，イ，ウ is perpendicular to Road エ.
○ Each of the Roads ア，イ，ウ is perpendicular to Road ク.
○ Each of the Roads ア，イ，ウ is perpendicular to Road ケ.
○ Each of the Roads ア，イ，ウ is perpendicular to Road コ.
(Note: the first four conditions are given at the beginning of the Park problem.)

Figure 3. Park, Problem Set B 5 (3), 2007 (translated by the author)

Table 2

Percentage of correct answers for the Park problem (Hirose, Nagashima, & Takahashi, 2014, p. 41)

Item	Percentage of correct answers (Total number of children)		
	Their children (23)	Tochigi prefecture (18 thousands)	National (1.1million)
(1)	95.8	73.0	71.2
(2)	83.3	79.0	79.2
(3)	12.5	13.8	17.9

Table 3

Percentage of the children's reasoning for finding the area of Central Park (Hirose et al., 2014, p. 42)

Children's reasoning for finding area of Central Park	Percentage
Choose height correctly and calculate the area of the parallelogram	12.5
Choose the slanted side as the height of the parallelogram	54.2
Miscalculation	4.2
Recall the wrong formula for the parallelogram	4.2
Calculate a different area	8.3
Calculate the circumference using addition	4.2
Use visual information	4.2
No response	8.3

4.2 *Lesson study using item (3) of the Park problem*

The observations regarding the children's responses to item (3) especially puzzled the teachers. Therefore, they decided to conduct a research lesson in a sixth grade classroom by modifying item (3) of the Park problem.

To develop the lesson plan, they developed some conjectures:

- We may need some devices to assist the children in understanding the task, in choosing the necessary information, and gaining perspective by linking the chosen information and the knowledge they have.

- The children may vary in their confidence in their knowledge and understanding.
- It would be important to modify and reflect on their own thinking and answers when they solve the problems.

About one month was spent on developing the lesson plan. The final lesson plan is shown in Figure 4 (Hirose et al., 2014, p. 52). The research lesson was conducted on February 7 with the 23 children who had been assessed. The lesson was video recorded and a post-lesson discussion was also conducted. To examine the lesson effect, the teachers collected the children's data prior, during, and after the lesson. Several children were chosen as focus students and their activities were recorded during the lesson using voice recorders and field memos.

Situation	Learning activity	Minute	Notice in teaching (* indicate connections with research theme)	Evaluation
Grasp/ Foresee	1 Grasp the task "Which of the geometrical figures is larger?"	5	* Show a figure that does not contain all information necessary to solve the task. - Make sure that A is parallelogram and B is rectangle. * Make them discuss/think about which seems larger to raise their motivation. - Present the task and through discussion, help them understand that they need to calculate the area to solve the task.	
Solve	2 Solve the task.	7	* Distribute worksheet that contains necessary information. Children think individually. - For those who had difficulty, give advice during between-desk instruction.	
	3 Conduct a questionnaire.	3	* Purpose of questionnaire is so that children reflect on their own thinking during the activity.	

Cultivate	4 Discuss the height. (1)Small group discussion (2)Whole-class discussion	10 10	* Present two methods for the calculation of the parallelogram. - Let them discuss in groups the reason for the calculation of each method. * Children present what they discussed. Prepare the manipulatives for use when necessary. - Through presentations and small group discussions, make sure of the height of the parallelogram.	
Cultivate	5 Summarize the lesson	10	* Encourage children to reflect and think about the area of the parallelogram, the height of which is sticking out of the base, by solving the task again and summarizing observations about the height. - Let them write their reflections on the worksheet	Mathematical thinking Worksheet Observation

The figure used in Learning activity 1

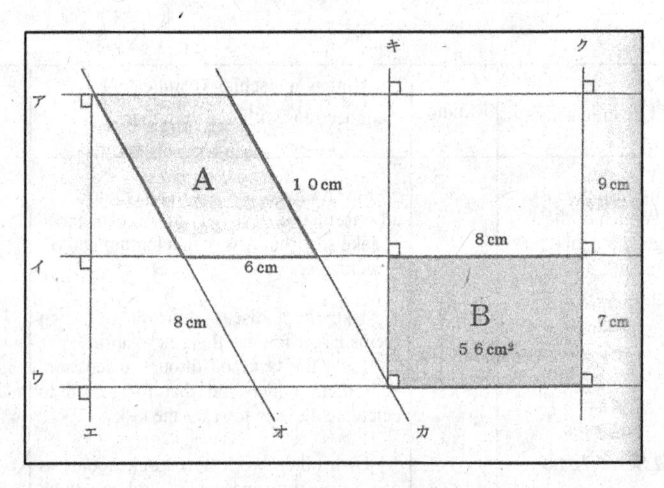

Two methods for calculation of the parallelogram presented to children in Learning activity 4

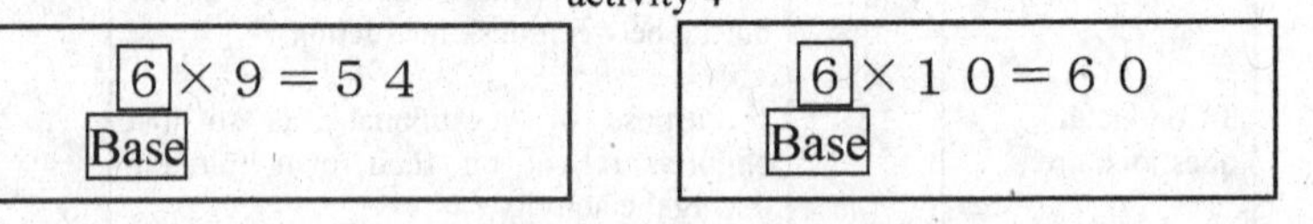

Figure 4. A lesson plan using item (3), Park problem (translated by the author)

Table 4 shows the percentage of correct answers for item (3) during and after the lesson. The percentage during the lesson was much higher than 12.5 (Table 2), which may be partly because the task was modified in the lesson, or because the children had some familiarity with the task. After the lesson, the percentage of correct answers had increased further.

During the lesson, the teacher conducted a questionnaire with the children (see Learning activity 3 in Figure 4). Here, the children were asked "what they were careful about when solving the task" and had to circle all the relevant items from a list of six items (Table 5). The same questionnaire was conducted again after the lesson when they had solved the task.

Table 4

Percentage of correct answers for item (3) during and after the lesson

Time	Percentage of correct answer
During the lesson	56.5
Three days after the lesson (item (3))	70.8
One week after the lesson (modified version of (3))	75.0

Table 5

Item and number of children who circled the item

Item	During the lesson	After the lesson	
		Item (3)	Modified version of item (3)
Figure	15	24	24
Problem statement	4	0	2
Calculation	16	12	8
Answer	8	5	7
Unit of measure	4	3	4
Other	1	0	0

Table 5 shows the items in the list and the number of children who circled each item. It is interesting to note that during the lesson, the items "figure" and "calculation" were the two major objects that the children were careful about. A further classification of the children regarding the items "figure" and "problem statement" revealed that there was a strong bias toward the relationship between the number of children who obtained the correct answer and the number of children who replied that they were careful about the figure and/or problem statements. Indeed, 11 of the 15 children and 3 of the 4 children who circled "figure" and "problem statement" respectively were the children who had the correct answers. On the other hand, many of the children who made mistakes circled the items "calculation" and "answer". Therefore, it could be conjectured that the children who made the mistake of choosing the slanted side tended to pay attention to the calculation but not to the map or the problem statement.

After the lesson, the number of children who circled "figure" increased greatly. Classroom observations of the children also showed that the children were talking about the superficial attention to the map with respect to the solution "6 × 10 = 60". For example, there were utterances among children during the small group discussion, "[someone who wrote 6 × 10 = 60 was] not looking at the figure carefully. He just made a guess" or "the person only looks at the very narrow area of the figure." From these data, it can be speculated that during the discussion about and the summary of the lesson, the children learned that they needed to pay careful attention to the figures so as to find the relevant information.

4.3 *The three teachers' reflections*

The teachers' final report (Hirose et al., 2014) contained their observations and reflections on the activities they were engaged in over the six-month period. First, the teachers felt that there were different stages in a child's state of knowledge (Table 6) and that these stages could be used to examine the research lesson effect (Hirose et al., 2014, p. 106). They further pointed out the importance of monitoring one's thinking and distinguished distinct stages. These descriptions indicated that they had a better perspective on a child's ability to utilize knowledge and skills; that is, the extent to which the child was conscious of the selection and the connections to their knowledge in the task they faced.

Second, the teachers felt that to improve a state of knowledge stage, it was effective to narrow the discussion focus and clarify the reasons for the decision making in the lesson (Hirose et al., 2014, p. 107). However, narrowing the discussion topic in small groups or in the whole class appeared to be a challenge for the teachers. Initially, they attempted to equally foster many aspects of the children's ability to utilize their knowledge and skills using five situations (see Figure 4) in one lesson. During the meeting, the rationale and purpose of setting all these situations were discussed. We finally agreed to focus on the "cultivating" situation, because this had the possibility of fostering the ability to evaluate and modify a student's own method. Many students had demonstrated difficulty in this ability; therefore, this situation was stressed in the final flow of the lesson.

Table 6
State of knowledge stages (Hirose et al., 2014, p. 106, translated by the author)

State of Knowledge	Descriptions
K-stage 4	To solve the task, the student can think by connecting different knowledge or choosing the necessary knowledge and can explain the thinking process to others.
K-stage 3	To solve the task, there remains some vagueness on connecting the different knowledge or choosing the necessary knowledge. They are unable to explain the thinking process to others but can solve the task on their own.
K-stage 2	To solve the task, there remains some vagueness on connecting the different knowledge or choosing the necessary knowledge. They are able to solve the task on their own sometimes.
K-stage 1	To solve the task, there remains some vagueness on connecting the different knowledge or choosing the necessary knowledge. They are unable to solve the task.

4.4 *Summary*

For the three practicing teachers, Problem Set B had many well-designed problems that targeted an ability to utilize knowledge and skills to solve the problems, which was the teachers' main research theme as their lesson studies were focused on one problem item from Problem Set B.

This item requires children to arrange the given information and to properly choose the relevant information. Therefore, it is necessary for the children to regulate their thinking by repeatedly checking whether the information they are choosing is relevant to the task at hand. The teachers were surprised at the low national average results; however, they were more surprised at the similarly low results of the children in their own classes.

The teachers' explorations could be seen to be an endeavor to listen more critically to the children's actual (underdeveloped) reflective thoughts and to respond to the children so as to assist them in deepening their understanding. Through the activities related to their research theme, the teachers gained a perspective as to how to develop the children's ability to think reflectively. Further, as they had the opportunity to conduct a lesson which focused on reflection, they were given the opportunity to re-conceptualize the lesson (see Hino & Makino, 2015, to read more about these teachers).

5 Listening and Responding to Children's Reflective Thinking

These two teacher education case studies illustrate the teachers' struggles in listening and responding to the children's often underdeveloped reflective thinking. In these case studies, listening and responding to the children's thinking were provided in the lesson design context because lesson design was considered fundamental to the development of the expertise required in both teacher preparation and professional development programs.

The relationship between the 21st century competencies which are generic in nature and the academic abilities that are more specific to each school subject has been discussed (e.g., MEXT, 2014; Shimizu, 2015). In this regard, the abilities related to "application" in the NAAA can be said to be the indicators that identify some of the competencies to be fostered in mathematics, of which reflective thinking is an essential part. Nevertheless, the case studies demonstrated that the teachers were less conscious of children's reflective thinking. In the case studies, one of the main problems highlighted was that the teachers tended to think that the children could solve the problems and explain their thinking with less difficulty. This finding implies that many teachers may overlook or ignore the children's reflective thinking while teaching mathematics. It was seen that this situation could be worse when teachers only followed the textbook exercises. To raise teacher awareness of children's thought processes, it is critical to develop mathematical problems that pay special attention to the targeting of reflective thinking skills. In the same line of thought, strategies for modifying textbook problems to incorporate higher-level cognitive demand are important for teachers (e.g., Kaur & Yeap, 2009a, 2009b).

To adequately respond to children's reflective thinking, it is important to elicit, attend to and make this thinking the object of the discussion (Cobb, Boufi, McClain, & Whitenack, 1997). However, the prospective teachers who were studied in this chapter tended to concentrate on devising scaffolds to circumvent the children's erroneous thinking, which suggests that the teachers' ways of listening and responding to the children's thinking were essentially *evaluative* (Davis, 1997). Further, the prospective teachers tended to anticipate the children's thinking based on their own experience in solving the same problems, which could be seen to be a feature of the initial teacher training (Takahashi, 2011). In this regard, the lesson ideas generated from the NAAA problems were effective as they provided the prospective teachers with the opportunity to recognize different ways to organize lessons to respond to the children's underdeveloped thinking by comparing them with their own methods.

Three practicing teachers conducted a lesson study in which listening and responding to children's reflective thinking was the center of the inquiry. However, this was not an easy task. When designing the lesson, teachers had difficulty in the design of the *neriage*, where negotiated, participatory and interactive actions (*hermeneutic listening*, Davis, 1997) were required. The teachers fully recognized the importance of *neriage*; however, they struggled with how to orchestrate the children's reflective thinking to assist them in honing their thinking in the last activity, which required a highlighting and summarizing of the major points (refer to Stigler & Hiebert, 1999, for the sequence of five activities in the structured problem solving). From these results, it appears that richer image of mathematics lessons are needed so that teachers can successfully foster children's reflective thinking.

6 Conclusion

It has been noted that Japanese lessons reflect certain societal values, one of which is self-reflectivity (e.g., Lewis, 1996; Stigler & Hiebert, 1999; Shimizu, 2009). Being based on such a value, the problem solving approaches in Japan include a powerful methodology for enhancing reflection (Hino, 2013). However, it is not sufficient to just follow the structured problem-solving activities. To successfully foster children's reflective thinking, teachers must pay attention to the children's real and often underdeveloped reflective thinking. Therefore, alongside having the knowledge of all aspects of teaching mathematics, teachers must also recognize the value of promoting reflective thinking and must have positive beliefs about the children's reflective thinking. Obviously, this requires teachers to be flexible while developing their lesson plans. Furthermore, this chapter has demonstrated that teacher abilities in listening and responding to children's reflective thinking can develop if they have the opportunity to study these ideas more closely. Collaborative research between teachers and teacher educators can offer invaluable information for both parties to develop expertise in fostering children's 21st competencies in the mathematics classroom.

References

Ball, D. L., Thames, M. H., & Phelps, G. (2008). Content knowledge for teaching: What makes it special? *Journal of Teacher Education, 59*(5), 389-407.

Beswick, K. (2013). Knowledge and beliefs for nurturing reflective learners of rational number concepts. In B. Kaur (Ed.), *Nurturing reflective learners in mathematics* (pp. 57- 79). Singapore: World Scientific.

Cobb, P., Boufi, A., McClain, K., & Whitenack, J. (1997). Reflective discourse and collective reflection. *Journal for Research in Mathematics Education, 28*(3), 258–277.

Crespo, S. (2000). Seeing more than right and wrong answers: Prospective teachers' interpretations of students' mathematical work. *Journal of Mathematics Teacher Education, 3*, 155-181.

Davis, B. (1997). Listening for differences: An evolving conception of mathematics teaching. *Journal for Research in Mathematics Education, 28*(3), 355–376.

Griffin, P., McGaw, B, & Care, E. (2012). *Assessment and teaching of 21st century skills*. Netherlands: Springer.

Hino, K. (2013). Mathematics lessons stimulating reflective learning: Japanese perspective. In B. Kaur (Ed.), *Nurturing reflective learners in mathematics* (pp. 247-268). Singapore: World Scientific.

Hino, K., & Makino, T. (2015). Mid-career teacher learning through collaboratively framed mathematics lessons. In C. Vistro-Yu (Ed.), *Proceedings of The 7th ICMI-East Asia Regional Conference on Mathematics Education* (pp. 313-320). Cebu City: Philippines.

Hirose, Y., Nagashima, H., & Takahashi, T. (2014). *Devising lesson fostering the ability of utilizing learned knowledge in mathematics.* Report of inservice teacher training, Tochigi Prefecture. (in Japanese)

IMPULS. (2015). *2012 Grade 6 Mathematics Set B*. Retrieved 31 December, 2015 http://www.impuls-tgu.org/en/cms/uploads/File/resource/2012_Grade_6_Mathematics_Set_B.pdf

Kaur, B. (2013). Nurturing reflective learners in mathematics: An introduction. In B. Kaur (Ed.), *Nurturing reflective learners in mathematics* (pp. 1-11). Singapore: World Scientific.

Kaur, B., & Yeap, B. H. (2009a). *Pathways to reasoning and communication in the primary school mathematics classroom*. Singapore: National Institute of Education.

Kaur, B., & Yeap, B. H. (2009b). *Pathways to reasoning and communication in the secondary school mathematics classroom*. Singapore: National Institute of Education.

Lewis, C. (1996). Fostering social and intellectual development: The roots of Japanese educational success. In T. Rohlen, & G. LeTendre (Eds.), *Teaching and learning in Japan*. New York: Cambridge University Press.

MEXT (2008). *Elementary school teaching guide for the Japanese course of study: Mathematics*. (English translation was carried out by the Asia-Pacific Mathematics and Science Education Collaborative at DePaul University in Chicago, Illinois, U.S.A., under contract from the U.S. Department of Education)

MEXT (2014). *Committee meetings on the educational aim, content, and assessment on the basis of qualities and abilities to be fostered: Points of discussion.* (in Japanese)

NIER (2007). *Report of 2007 National Assessment of Academic Ability, Elementary School.* Retrieved 31 December, 2015 from http://www.nier.go.jp/tyousakekka/gaiyou_shou/19shou_houkoku4_2.pdf (in Japanese)

NIER (2008). *Report of 2008 National Assessment of Academic Ability, Elementary School.* Retrieved 31 December, 2015 from https://www.nier.go.jp/08chousakekka houkoku/08shou_data/houkokusho/05_shou_bunseki_sansuu.pdf (in Japanese)

NIER (2012). *Report of 2012 National Assessment of Academic Ability, Elementary School.* Retrieved 31 December, 2015 from https://www.nier.go.jp/12chousakekka houkoku/03shougaiyou/24_shou_houkokusyo-4_sansuub.pdf (in Japanese)

NIER (2013). *Fundamental principles on the organization of curriculum fostering abilities that correspond to the change of society* (Report No. 5, Basic Research on Curriculum Organization) (in Japanese)

NIER (2015a). *Manual of 2015 National Assessment of Academic Ability, Elementary School.* Retrieved 31 December, 2015 from http://www.nier.go.jp/15chousa/pdf/15kaisetsu_shou_sansuu.pdf (in Japanese)

NIER (2015b). *Examples of ideas for lesson on the basis of the results of National Assessment of Academic Ability.* Retrieved 31 December, 2015 from http://www.nier.go.jp/jugyourei/index.htm (in Japanese)

OECD (2005). *The definition and selection of key competencies: Executive summary.* Retrieved 31 December, 2015 from http://www.oecd.org/pisa/35070367.pdf

Polya, G. (1985). *How to solve it: A new aspect of mathematical method* (2nd ed.). Princeton and Oxford: Princeton University Press. (Original work published 1945)

Rodgers, C. (2002). Defining reflection: Another look at John Dewey and reflective thinking. *Teachers College Record, 104*(4), 842-866.

Shimizu, Y. (2009). Characterizing exemplary mathematics instruction in Japanese classrooms from the learner's perspective. *ZDM Mathematics Education, 41*, 311-318.

Shimizu, Y. (2015). The place of "mathematical methods" in the mathematics curriculum. *Proceedings of the Third Spring Conference of Japan Society of Mathematical Education*, 173-178. (in Japanese)

Stigler, J. W., & Hiebert, J. (1999). *The teaching gap.* New York: Free Press.

Takahashi, A. (2011). The Japanese approach to developing expertise in using the textbook to teach mathematics. In Y. Li, & G. Kaiser (Eds.), *Expertise in mathematics instruction: An international perspective* (pp. 197-219). New York: Springer.

Chapter 8

Using Open-Ended Tasks to Foster 21st Century Learners at the Primary Level

YEO Kai Kow Joseph

There are many possible solution approaches and strategies to solve an open-ended task. The implementation of open-ended tasks in the classroom provides opportunities to develop our pupils' 21st century competencies. This chapter therefore reviews the concept of open-ended tasks, discusses research studies related to open-ended tasks, deliberates two types of open-ended tasks and examines the issues in implementing open-ended tasks. Two types of open-ended tasks are highlighted so that teachers might trial in their mathematics lessons to develop pupils' decision making, thinking and reasoning process.

1 Introduction

The need to acquire 21st century skills is of international concern and interest. Innovation and new habits of thinking are critical for students to thrive in an increasingly diverse and complex setting. Access to and user-friendliness of new technology have facilitated individuals to push boundaries, and it thus comes as no surprise that today's students are able to "solve problems that involve multifaceted solutions, encounter issues that tests their values, and face challenges that are not documented in manuals and textbooks" (Liu & Tan, 2015, p. 336). Moreover, it is important that 21st century learners acquire an understanding of mathematical concepts, competency with thinking skills, and a positive attitude in learning. Learning mathematics is about connecting ideas to each other, seeking insights into those ideas, and solving problems that involve multiple steps and approaches. The mathematics curriculum in Singapore is consistent with 21st century reform-based visions of schooling around the world, such as the NCTM standards (NCTM 2000). The newest mathematics syllabus in Singapore (Ministry of Education, 2012), which was released in 2012 and implemented in 2013, continued to maintain mathematical problem solving as the central focus and listed three aims for the Primary Mathematics Syllabus. The three aims are to enable all pupils to:

- acquire mathematical concepts and skills for everyday use and continuous learning in mathematics;
- develop thinking, reasoning, communication, application and metacognitive skills through a mathematical approach to problem solving;
- build confidence and foster interest in mathematics. (p. 8)

The primary mathematics curriculum aims are to meet the demands of the 21st century for all pupils. To meet these demands, mathematics teachers need to be aware that 21st century skills are not a discrete set of skills to be taught over a short period of time or a set of skills for older pupils. It must be part of primary school pupils' daily mathematical lessons that slowly grows and develops over time. In addition, we need to educate pupils to have a robust mathematical knowledge while at the same time teaching them 21st century life skills which include the abilities to be synthesisers of information, good decision makers, good communicators and creative learners.

Mathematics educators, including school teachers, are now beginning to pay attention to the kind of tasks they give to their pupils. The Singapore mathematics curriculum has advocated that problems should cover a wide range of situations from routine mathematical problems to problems in unfamiliar context and open-ended tasks that apply the relevant mathematics knowledge and thinking processes (Ministry of Education, 2000). Success in problem solving is related to pupils' disposition and monitoring of their own thinking processes. It is an effective problem-solving experience if pupils make their own decisions on how to solve open-ended tasks rather than to follow taught procedures. To solve open-ended tasks pupils must observe, relate, question, reason and infer as well as employ various forms of representations. This chapter therefore reviews the concept of open-ended tasks, discusses research studies related to open-ended tasks, illustrates two types of open-ended tasks and examines the issues in implementing open-ended tasks.

2 Review of Literature

This section explains what constitutes an open-ended task and the benefits of using open-ended task in the classroom. In addition, research studies on pupils' solving open-ended tasks are reviewed.

2.1 *Open-ended tasks*

Although there are various views of open-ended tasks to be used in the school mathematics instruction, we can still identify some basic characteristics of open-ended tasks. Open-ended tasks "have more than one answer and/or can be solved in a variety of ways" (Moon & Schulman 1995, p. 25). Hancock (1995) also agreed that open-ended tasks could be considered to have more than a single correct solution and that they offer pupils multiple approaches to the problems by placing little constraints on the pupils' methods of solution. Similarly, Takahashi (2000) also shared that there are two types of open-ended tasks: problems with only one solution but diverse approaches and problems with multiple correct answers. In addition to producing a solution, pupils must also explain their solution process and justify their answer. In the earlier 1970s, Shimada with other researchers provided an incomplete problem which does not define clearly what the question asks for, therefore allowing many possible solutions (Becker & Shimada, 1997). Shimada and his research teams identified such a problem as an open-ended problem, and the approach adopted was called "open approach" or "the open-ended approach". Open-ended approach teaching has been used in Japan since the 1970s in order to stimulate higher-order thinking in mathematics (Becker & Shimada, 1997; Nohda, 1986). The idea of the open-ended approach in teaching mathematics is to encourage pupils to use their natural ways of mathematical thinking and to apply their previously acquired knowledge to the process of solving open-ended tasks (Kwon, Park, & Park, 2006). The open-ended approach was also a pedagogical strategy that intended to foster creative mathematics actions that arouse the pupils' curiosity when solving the problems.

Christiansen and Walther (1986) believed that open-ended tasks offer ideal conditions for cognitive development in which new knowledge is generated and ideas of earlier attained knowledge are applied and assessed. Effective uses of open-ended tasks are believed to foster higher-order thinking and promote, produce, and provide fodder for pondering (Dyer & Moynihan, 2000). According to De Lange (1995), a task that is open for pupils' process and solution is a way of motivating pupils' high quality thinking. Caroll (1999) added that short, open-ended tasks create an opportunity for the mathematics teachers to have quick checks into their pupils' thinking and conceptual understanding. They are no more time-consuming to correct than the exercises that teachers usually give. In the same vein, Kwon, Park and Park (2006) also reported that divergent thinking in mathematics could be accomplished through an open-ended approach because open-ended tasks show numerous explanations and the ways pupils solve the tasks based on their own choice and range of abilities. Furthermore, Sullivan and Lilburn (2005) explained that open-ended tasks are exemplars of good questions in that they advance significantly beyond the surface. The openness of good questions offers many teaching moments for classroom teachers because of their potential for pupils at different stages of development to respond at their own level (Sullivan, Clarke, & Wallbridge, 1991). Osana, Lacroix, Tucker and Desrosiers (2006) expressed that the use of open-ended tasks favours pupils' involvement in class activities and encourage them to explore and investigate. They also revealed that the use of open-ended tasks increase the pupils' motivation to generalise, look for models and links, communicate, discuss and identify alternatives. To summarise, open-ended tasks are those that require pupils to think more intensely and to provide a solution which involves more than remembering a fact or repeating a skill. There are many possible solution approaches and strategies to solve an open-ended task. The focus of open-ended tasks offer opportunities for pupils to reveal their decision making process, mathematical thinking, reasoning process as well as problem solving and communication skills.

2.2 *Research studies on open-ended tasks*

Earlier research studies by Sullivan and Clark (1992) studied the quality of responses given by the pupils to open-ended but content specific questions. They found that pupils were able to give multiple correct responses to such questions. Pehkonen (1995) conducted an experiment to explain the effects of open-ended tasks on the pupils' motivation towards Mathematics. The results showed that the experiment using open-ended tasks have significant positive effects on the pupils' motivation towards Mathematics. Van den Heuvel-Panhuizen (1996) explored how pupils responded to realistic problems when incomplete information was given. The pupils were required to make their assumptions on the missing information. Van den Heuvel-Panhuizen concluded that the use of open-ended approach provided the teacher with substantial evidence on how the pupils managed the problem-solving process. In international research studies on U.S. and Asian students' mathematical problem solving using open-ended tasks, it was found that there was a noticeable difference between U.S. and Asian students' solution representations (Becker, Sawada, & Shimizu, 1999; Cai, 1995; Cai, 2000; Silver, Leung, & Cai, 1995). The studies showed that Asian students inclined to use symbolic representations (e.g., arithmetic or algebraic symbols) while U.S. students inclined to show visual representations (e.g. diagrams). Klavir and Hershkovitz (2008) showed that creative mathematical thinking could be nurtured by providing open-ended tasks as an assessment tool for both teachers and students.

Local studies on the use of open-ended tasks in mathematics lessons too revealed positive outcomes. Seoh (2002) explored how an Open-ended Approach (OEA) mathematics programme was used to develop critical thinking skill in mathematics of 52 pupils from Secondary 5 (17 year-olds) normal academic stream in Singapore. The study, implemented for eight weeks, comprised eight problem-solving lessons that integrated the use of open-ended tasks and 'working in pairs' to encourage the pupils to think and reason critically. It was found that the OEA programme enhanced the pupils' critical thinking skill in mathematics. After the implementation of the OEA, the pupils' ego of wanting to outperform their classmates had reduced. They were more willing to engage in cooperative learning with their classmates and more willing to help their fellow weaker classmates to improve their mathematics performance together. In addition, their avoidance towards solving mathematical problems had improved. They were keen to work on their mathematics and did not avoid work associated with mathematics. On the whole, the pupil who underwent the study gave positive feedbacks towards the Open-ended Approach programme.

3 Types of Open-Ended Tasks

The brief review of literature in the previous section has provided a sense of what open-ended tasks can be intended for. With the introduction of the 2013 Mathematics syllabus in Singapore, it is appropriate to revisit open-ended tasks in terms of their relevance and applicability to the teaching of primary mathematics. While the Singapore revised mathematics syllabus (Ministry of Education, 2012) continues to emphasise learning experiences in the classroom, there is now an even greater emphasis on the development of pupils' 21st century skills such as decision making and communication. Anecdotal evidence from many experienced mathematics teachers suggests the development of pupils' 21st century skills is best if the focus is consistent and happens regularly during mathematics lesson. Moreover, mathematics teachers need to have a well-developed and meaningful understanding of open-ended tasks themselves and also pedagogical practices that provide learning experiences for pupils to explore and construct multiple solutions for open-ended tasks. Selecting appropriate open-ended tasks depends on the instructional objectives of the lesson.

Despite the benefits of open-ended tasks, it is a challenge to formulate different types of open-ended tasks. For this reason, the tasks with the following two features will also be regarded as open-ended tasks in this chapter. First, the initial point of the task is rather clear but the solutions for its objectives can vary. Second, they are tasks in which pupils can show higher-order thinking skills and employ divergent thinking in the search of their own solutions. The focus of this chapter will be on short open-ended tasks that teachers can adapt from closed questions found in textbook and workbook exercises. Teachers can use short, open-ended tasks for their role in teaching through problem solving that emphasises learning mathematical concepts and skills through a problem situation. The following section exemplifies open-ended tasks that make assumptions on the missing information and open-ended tasks that discuss a concept, algorithm or error as two possible categories that can be incorporated in the teaching and learning of mathematics at the primary level.

3.1 *Open-ended tasks that make assumptions on the missing information*

This type of open-ended task requires pupils' own contribution to the process such as making assumptions on the missing information. In addition, pupils also need to solve the problem where there is no known solution beforehand and not all data are given. The content specific open-ended tasks in this category can be illustrated by means of some examples below.

- Draw a triangle. Write a number in the centre of the triangle. Write three numbers in the corners of the triangle that add up to the number in the centre. Now challenge yourself by choosing greater numbers.
- List five 3-digit numbers that have the digit 7 in the tens place.
- Draw a rectangle where the area of the rectangle is $24cm^2$.
- Write five decimals between 10 and 20.
- Find the dimensions of 2 different boxes so that each has a volume of $150cm^3$.

The open-ended task 1 (see Figure 1) aims to stimulate Primary 1 pupils' high-level thinking processes and mathematical creativity. In open-ended task 1, pupils, in addition to applying addition and subtraction algorithms, were also required to solve problems where there are no known or fixed procedures and not all information is given. It would require pupils' own contributions, such as making assumptions on the missing data. As the numbers of green balls or yellow balls to be taken out from the box are not provided, pupils have to think about and decide the number for each colour of the balls. There is no cue word for pupils to figure out which operation to use, as in a one-step word problem.

Open-Ended Task 1 (Primary 1)
There are 6 green balls and 6 yellow balls in a box.
I take out any 7 balls from the box.
How many of them are green balls?
How many of them are yellow balls?
Explain clearly using pictures, numbers and words.

Figure 1. Open-ended task 1

From the curriculum perspectives, this open-ended task 1 has been designed specifically for the following cognitive demands:

- Pupils identify missing information essential to the open-ended task.
- Pupils make their own assumptions about the missing information: number of green balls and yellow balls to be taken out.
- Pupils access relevant mathematical concepts of addition and subtraction of one-digit and two-digit whole numbers.
- Pupils work out the number bonds and operations.
- Pupils must communicate their mathematical reasoning in words and pictures.
- Pupils show their creativity in using possible heuristics and solutions.

Figure 2 and Figure 3 show the responses of two primary 1 pupils. The uses of verbal, symbolic and pictorial mathematical representations were manifested in these children's work when they were given the opportunity to be creative in an open-ended situation such as this.

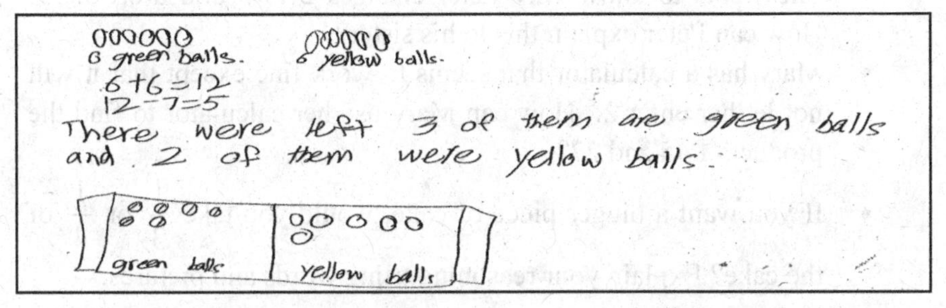

Figure 2. Pupil A's responses to open-ended task 1

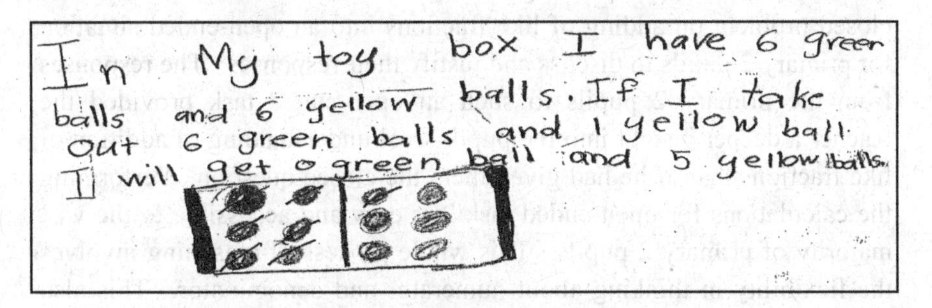

Figure 3. Pupil B's responses to open-ended task 1

3.2 *Open-ended tasks to discuss a Concept, Algorithm or Error*

In a traditional classroom, primary school pupils have little opportunity to discuss and justify the mathematical process involved in their mathematical solutions. Sometimes they may not be able to explain their thinking. Although they may perform certain computations, they do not know how to discuss how they solve them or why they work. Even if the teacher insists that the pupils discuss and justify, they may simply mimic what the teacher has said in class. The content specific open-ended tasks in this category can be illustrated by means of some examples below.

- You cannot remember what is 9 x 7 but you remembered that 5 x 7 is 35. Explain how could you use this fact to work out 9 x 7?
- Peter is helping his sister to round off 8.29 to the nearest tenth. She wants to know why Peter changes 2 to 3 and drop the 9. How can Peter explain this to his sister?
- Mary has a calculator that seems to work fine except that it will not let her enter 2. How can Mary use her calculator to find the product of 45 and 27?
- If you want a bigger piece of cake, would you take $\frac{1}{3}$ or $\frac{1}{4}$ of the cake? Explain your reasoning using words and pictures.

Open-ended task (see Figure 4) was converted from a standard closed problem on adding of like fractions into an open-ended situation for primary 2 pupils to discuss and justify their responses. The responses from the primary 2 pupils to such an open-ended task provided the teacher a deeper insight into the pupils' real understanding of addition of like fractions than if he had given them the closed question. Performing the calculations for open-ended task 2 is easy and accessible to the vast majority of primary 2 pupils. This whole process of reasoning involves the flexibility in thinking about numerator and denominator. This also creates an opportunity for pupils to explore and appreciate fractions as part of whole.

Open-Ended Task 2 (Primary 2)

John wrote $\frac{1}{6}+\frac{2}{6}=\frac{3}{12}$.

Explain clearly using words and picture to show John's answer cannot be correct.

Figure 4. Open-ended task 2

From the curriculum perspectives, this open-ended task 2 has been designed specifically for the following cognitive demands:

- Pupils identify the error showed in the task.
- Pupils access relevant mathematical concept to correct the error.
- Pupils must communicate their mathematical reasoning in words and pictures.
- Pupils show their creativity in using possible diagrams and solutions.

Figure 5 shows a good but not the best response. Pupil A is able to correct and show some form of mathematical reasoning but the generalisation is not explained adequately. Pupil A's response is partially correct as it is shown in the last two lines how three-sixths of the circle is shaded. Pupil B's responses (shown in Figure 6) would have been scored as correct in the closed question. However, it appears pupil B lacks the reasoning and communication skills as the justification is unclear. Their explanations show that both would need remedial assistance on the concept of addition of like fractions and how to represent them. Open-ended task to discuss errors could be a good platform for the teacher to engage pupils in extending their thinking and reasoning as there are several ways of justifying it.

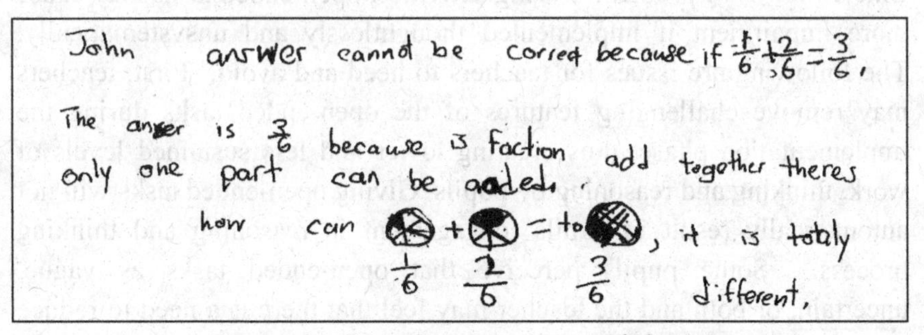

Figure 5. Pupil A's responses to open-ended task 2

$\frac{1}{6} + \frac{2}{6}$ cannot be equal to $\frac{3}{12}$ because:
It is fraction not add, the $\frac{3}{12}$ is correct
but the below one is not it should
be $\frac{3}{6}$.
The denominator is wrong also, thats why
the answer cant get write.

Figure 6. Pupil B's responses to open-ended task 2

These two different types of open-ended tasks exemplify how teachers and pupils could benefit from implementing open-ended tasks in the primary mathematics classroom. The two types of open-ended tasks are just a first step towards developing pupils' 21st century skills in the classroom to meaningful ones where the emphasis is on the process (reasoning, thinking and decision making) rather than the product (final answer).

4 Issues in Implementing Open-Ended Tasks in the Classroom

Like all forms of problem-solving activities, open-ended tasks may cause more impairment if implemented thoughtlessly and unsystematically. The following are issues for teachers to heed and avoid. First, teachers may remove challenging features of the open-ended tasks during the implementation phase, thus creating lower and less sustained levels of work, thinking and reasoning by pupils. Giving open-ended tasks will not automatically result in pupils' engagement in reasoning and thinking process. Some pupils perceive that open-ended tasks as vague, uncertain, or both and the teacher may feel that there is a need to reduce their complexity so as to lessen the pupils' anxiety towards such tasks. Reduction in complexity can happen in numerous ways, including through pupils' successfully pressuring the teacher to provide clear steps to work out the task or the teacher explains and shows the challenging aspects of the task. If these are accomplished for the pupils, the cognitive demands of the open-ended task are weakened and pupils' thinking processes becomes directed into more predictable and instrumental understanding.

Second, the classroom-based shift that moves away from meaning and understanding towards the completeness or correctness of answers should not be the priority when attempting open-ended task. Teachers need to be mindful that the anticipated outcome of the open-ended task is not just showed through the solution. It is also the cognitive processes involved in reaching the solution. Teachers must have a paradigm shift towards a more process-based approach.

Third, open-ended tasks may decline into predetermined method and answer if too little time is allocated for the pupils to attempt. In such scenarios, pupils may be deprived of the time needed to truly engage with the mathematics content as well as to explore and involve in mathematical thinking.

Fourth, the success of open-ended tasks is the consideration of the relationship between pupils and task. Teachers must know their pupils well in order to make right choices regarding the motivational appeal, difficulty level and degree of task explicitness. Nohda (1986) indicated that even pupils with a lower motivation level could be involved in solving open-ended tasks. Appealing to pupils' natural ways of thinking lies at the heart of the "open approach" as pupils will produce different solutions and then share their thinking with the rest of the class. This is to move pupils into the correct cognitive and affective space so that thinking and reasoning can emerge and progress can be made on the task.

Finally, if instructional procedures at the implementation phase to solve open-ended task are not articulated to the pupils, classroom management issues may surface. Pupils may be engaged in off-task activities. This suggests that teachers are struggling with keeping pupils under control in addition to keeping them focused on the mathematics. Anthony and Walshaw (2009), for example, in a research synthesis, concluded that "in the mathematics classroom, it is through tasks, more than in any other way, that opportunities to learn are made available to the students" (p. 96). The issues discussed have implications on the roles of the teacher. Not only must the teacher choose and appropriately set up worthwhile open-ended tasks but the teacher must also proactively and consistently support pupils' cognitive activity without reducing the difficulty and cognitive demands of the open-ended task.

5 Concluding Remarks

In the traditional approach, there has been a tendency for pupils to view mathematics as simply practicing one-step, two-step or many-step procedures to find answers to routine problems. On the contrary, open-ended tasks if presented on a regular basis would impart in them that understanding and explanation are critical aspects of mathematics. Using open-ended tasks effectively in the classroom requires effective facilitation on the part of the teacher. This is a different approach to teaching from what most teachers have experienced, and it requires time and practice to learn. Ultimately, the decision to use open-ended tasks in the mathematics lessons is up to the teacher. It is hoped that teachers will bear in mind the appropriate use of open-ended tasks by relating it to their pedagogical goals and their pupils' abilities.

References

Anthony, G., & Walshaw, M. (2009). Effective pedagogy in mathematics. *Educational Series—19.* Brussels: International Bureau of Education, Geneva.

Becker, J. P., Sawada, T., & Shimizu, Y. (1999). Some findings of the U.S.-Japan cross-cultural research on students' problem-solving behaviors. In G. Kaiser, E. Luna, & I. Huntley (Eds.), *International comparisons in mathematics education* (pp. 121–139). London: Falmer Press.

Becker, J. P., & Shimada, S. (Eds.) (1997). *The open-ended approach: A new proposal for teaching mathematics.* Reston, VA, National Council of Teachers of Mathematics.

Cai, J. (1995). A cognitive analysis of U.S. and Chinese students' mathematical performance on tasks involving computation, simple problem solving, and complex problem solving. *Journal for Research in Mathematics Education Monograph Series 7*. Reston, VA: National Council of Teachers of Mathematics.

Cai, J. (2000). Mathematical thinking involved in U.S. and Chinese students' solving process-constrained and process-open problems. *Mathematical Thinking and Learning*, 2, 309–340.

Carroll, W. M. (1999). Using short questions to develop and assess reasoning. In L.V. Stiff, & R. Curcio (Eds.) *Developing mathematical reasoning in grades K-12, 1999 Yearbook* (pp. 247- 255). Reston, Va.: NCTM.

Christiansen, B., & Walther, G. (1986). Task and activity. In B. Christiansen, A.G. Howson, & M. Otte (Eds.) *Perspectives on mathematics education* (pp. 243-307). Holland: D. Reidel.

De Lange, J. (1995). Assessment: No change without problems. In T. A. Romberg (Ed.) *Reform in school mathematics and authentic assessment* (pp. 87-173). New York: Suny Press.

Dyer, M., & Moynihan, C. (2000). *Open-ended question in elementary mathematics instruction & assessment.* Eye on Education.

Hancock, C.L. (1995). Enhancing mathematics learning with open-ended questions. *The Mathematics Teacher, 88*(6), 496-499.

Klavir, R., & Hershkovitz, S. (2008). Teaching and evaluating 'open-ended' problems. *International Journal for Mathematics Teaching and Learning, 20*(5), 23.

Kwon, O., Park, J., & Park, J. (2006). Cultivating divergent thinking in mathematics through an open-ended approach. *Asia Pacific Education Review, 7*(1), 51–61.

Liu, W. C., & Tan, O. S. (2015). Teacher effectiveness: Beyond results and accountability. In O. S. Tan & W. C. Liu (Eds.), *Teacher effectiveness: Capacity building in a complex learning era* (pp. 335–345). Singapore: Cengage Learning Asia

Ministry of Education (2000). *Mathematics syllabus: Primary*. Singapore: Curriculum Planning and Development Division.

Ministry of Education (2012). *Primary mathematics: Teaching and learning syllabus*. Singapore: Curriculum Planning and Development Division.

Moon, J., & Schulman, L. (1995). *Finding the connections: Linking assessment, instruction, and curriculum in elementary mathematics.* Portsmouth, N.H: Heinemann.

National Council of Teachers of Mathematics (2000). *The principles and standards for school mathematics.* Reston, VA: Author.

Nohda, N. (1986). A study of "open-approach" method in school mathematics. *Tsukuba Journal of Educational Study in Mathematics, 5,*119–131.

Osana, H., Lacroix, G., Tucker, B. J., & Desrosiers, C. (2006). The role of content knowledge and problem features on preservice teachers' appraisal of elementary tasks. *Journal of Mathematics Teacher Education, 9*(4), 347-380.

Pehkonen, E. (1995). Using open-ended problem in mathematics. *Zentralblatt fur Didaktik der Mathematik, 27*(2), 67-71.

Seoh, B. H. (2002). An open-ended approach to enhance critical thinking skill in mathematics among secondary five normal (academic) pupils. Unpublished master's dissertation, Nanyang Technological University, Singapore.

Silver, E. A., Leung, S. S., & Cai, J. (1995). Generating multiple solutions for a problem: A comparison of the responses of U.S. and Japanese students. *Educational Studies in Mathematics, 28*(1), 35–54.

Sullivan, P., Clarke, D.J., &: Wallbridge, M. (1991) *Problem solving with conventional mathematics content: Responses of pupils to open mathematical tasks. Mathematics* Teaching and Learning Centre Report No.1.

Sullivan, P., & Clarke, D.J. (1992). Problem solving with conventional mathematics content: Responses of pupils to open mathematical tasks. *Mathematics Education Research Journal* 4(1), 42-60.

Sullivan, P., & Lilburn, P. (2005). *Open-ended maths activities: Using 'good' questions to enhance learning.* Melbourne: Oxford University Press.

Takahashi, A. (2000). Open-ended problem solving enriched by the internet. Paper presented at the NCTM annual meeting, Chicago, IL. Retrieved 29 February, 2012, from http://www.mste.uiuc.edu/users/aki/open_ended/NCTM_Presentation/sld006.htm.

Van den Heuvel-Panhuizen, M. (1996). *Assessment and realistic mathematics education.* Utrecht: CD-B Press/Freudenthal Institute, Utrecht Univerisity.

Chapter 9

Productive Talk in the Primary Mathematics Classroom

KOAY Phong Lee

Learning is a social endeavor. Hence, talking with teachers and fellow students about ideas is fundamental to student construction of mathematical knowledge and achievement. Many teachers acknowledge the value of mathematics talk but have found the implementation of productive mathematics talk to be challenging. In this article, a rich task is used to illustrate how talk moves can be employed to engage students in discourse.

1 Communication

The fast changing world propelled by increasing globalization and technological advancements has led both educators and business leaders to reexamine the aim of education and identify the competencies required for school leavers so that they would lead a meaningful life in the 21st century. These competencies, particularly communication skills, are critical to the advancement of Singapore, a nation with scarce natural resources. Singapore is a business hub and Singaporeans need to be able to communicate clearly and effectively with their counterparts elsewhere while negotiating business deals and setting up business ventures. Another important sector in the Singapore economy is the service sector. It is essential for people working in this sector to be able to communicate effectively with customers and clients. Hence, communication is one of the 21st century competencies identified by the Singapore Ministry of Education as necessary for all students. The other competencies include civic literacy, global awareness and cross-cultural skills; critical and inventive thinking, collaboration and information skills (MOE, 2010).

Communication is the exchange of thoughts or knowledge between two or more people through speech, visual representations, signals, writing, or behavior. It is also about the understanding of the intentions underlying the information conveyed. Hence, effective communication is about how we convey information so that it is received and understood in exactly the way we intended. It is a skill that every student should demonstrate in all subjects including mathematics.

In the current Singapore mathematics syllabus documents, communication refers to "the ability to use mathematical language to express mathematical ideas and arguments precisely, concisely and logically. It helps students develop their understanding of mathematics and sharpen their mathematical thinking." (MOE, 2012, p. 17). Teachers are expected to include the use of discourse in their classrooms to provide a more engaging, student-centred learning environment that focuses on sense-making and problem solving. They are encouraged to provide opportunities for all students not only to do mathematics, but also to talk and write about mathematics. Students are expected to use appropriate mathematical language, notations, symbols and conventions to explain fluently, clearly and precisely, their mathematical reasoning and thinking. Moreover, they are to analyze and evaluate their friends' thinking and strategies, and reflect on their own learning. Even if the views expressed are different from their own, students should listen with empathy and try to understand what their friends are saying. Hopefully, students would then become confident problem solvers who are able to work collaboratively, and also independently, think critically and communicate effectively.

Communication skills in a mathematics classroom include writing skills, listening skills, talking skills, non-verbal communication skills and thinking skills (Figure 1). In order to have productive talk, we need to listen to what others are saying and responding, organize, process and expand on our thoughts before verbalizing them. Often, we also use writing (including drawing) and non-verbal communication (e.g. tone, gesture, posture and facial expression) to put our ideas across. This article addresses the use of classroom talk to promote mathematics learning as talking about ideas is fundamental to learning.

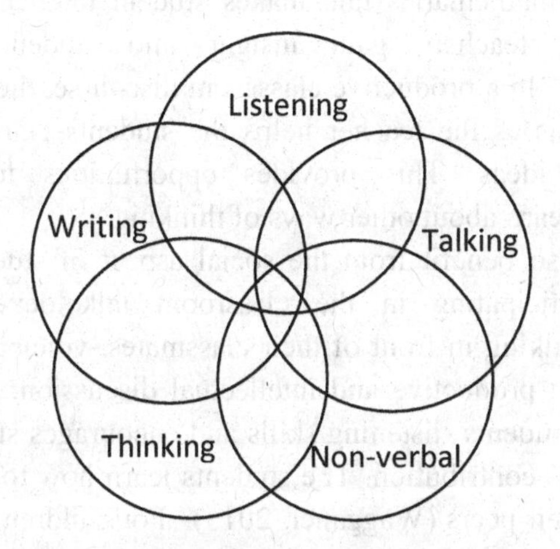

Figure 1. Communication skills (mathematics)

2 Purpose of Mathematics Talk

In many mathematics classrooms in Singapore, teachers are often the main source of the mathematics ideas and are responsible for the student learning. They dominate the classroom talk and are usually the sole questioners. The questions can broadly be categorized as performative, procedural and conceptual. The student contribution in the class discussion is often limited to answering of questions addressed to them and when asked, sharing of their solutions, describing or stating a sequence of procedures they have used to solve a problem, with little or no response from the other students (Hogan, Rahim, Chan, Kwek, & Towndrow, 2012; Hogan, 2014).

Mathematics talk in the classroom has other benefits besides imparting knowledge and assessment. In a classroom where students have opportunities to develop and express their ideas and initiate questions, solve problems, explain and justify their solutions, they are actively constructing their mathematical knowledge and thinking like mathematicians, engaging in verbal conjecture, defending and modifying their thoughts. Such discussions of mathematical ideas foster reasoning and help students develop mathematical understanding (NCTM 2000).

Moreover, mathematics talk makes student mathematical thinking visible, letting teachers gain insight into student thought and misconceptions. In a productive classroom discourse, the feedback is on an ongoing basis as the teacher helps the students clarify, extend and improve their ideas. This provides opportunities for students to contribute and learn about other ways of thinking.

Students also benefit from the social aspect of student talk in the classroom. Participating in the classroom talk develops students' confidence in talking in front of their classmates, voicing their thoughts and engaging in productive and intellectual discussion. Classroom talk also improves students' listening skills and encourages students to value their classmates' contribution. The students learn how to listen and how to respond to their peers (Wagganer, 2015). For children, classroom talk helps them develop communication skills such as looking at the person they are talking with, taking turns when communicating and knowing how to compromise and solve verbal conflicts (Seefeldt, 2004).

3 Features of Effective Mathematics Talk

Communication in the mathematics classroom occurs when students represent, argue, exchange, explain and defend their ideas. Mathematics talk can occur as whole class discussion, small group discussion, partner talk or student presentation. It is shaped by the tasks given, language choices in the classroom and the classroom environment.

For productive classroom talk, the teacher must first establish a warm and non-competitive learning environment where students listen attentively to and respect one another. Students feel free and are not afraid to voice their thoughts knowing that their ideas will be considered and not be ridiculed. In addition, the talk is not dominated by a handful of students (Clark, Jacobs, Pittman, & Borko, 2005; Michaels, O'Connor, Hall, & Resnick, 2013).

According to the Cockcroft report (1982, para. 306), "language plays an essential part in the formulation and expression of mathematical ideas" and there is a need to "extend and refine the use of mathematical language in the classroom." The language of mathematics is complex. It involves mathematical notations and rules as well as the vocabulary. The use of correct language of mathematics helps students to construct correct mathematical concepts and communicate them properly and effectively. Often students are able to read word problems in English text but are not able to transform them into mathematical equations or sentences if they have not fully understood the language of mathematics as shown in the case below (Figure 2). The child was able to identify key words such as 'fewer than' and 'less than' and knew that the two unequal sets were being compared. This knowledge did not necessarily lead to the translation to the correct mathematical equation.

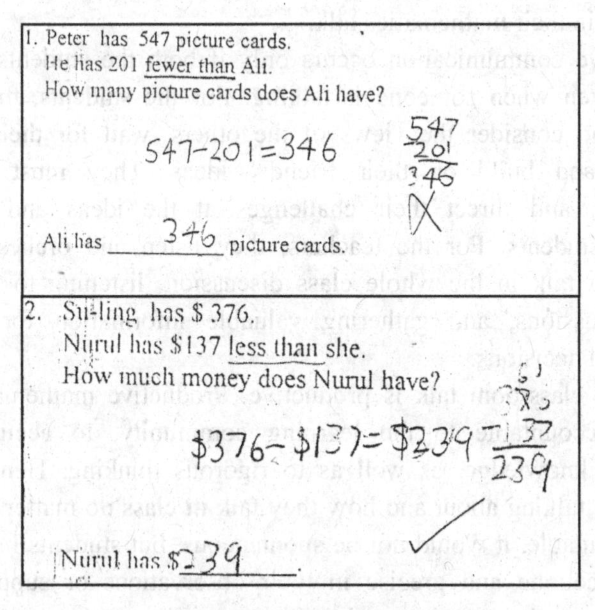

Figure 2. Sample of student's work

In the follow-up discussion with the student, the teacher may ask if Peter has fewer picture cards than Ali, then who has more cards, Peter or Ali? Such questions devoid of numbers would lead students to focus on the relationship rather than on which operation to use to compute the answer.

Communication in mathematics, whether written or verbal, centres upon the proper, correct and effective use of the language of mathematics. In the classroom, in order to participate in mathematics talk, students must understand and use the language of mathematics, and the teachers must explicitly teach the language, use it themselves and require their students to do so. For example, when talking about the product of two numbers, the students may say that the product of two numbers is greater than the factors and may forget to add that it is true only with whole numbers greater than one. As the teachers press for clarification and explanation, they model the thought processes by probing and questioning while using appropriate language. Gradually, students can be expected to develop such habits of mind in pair or group discussions. With frequent practice, students will become more confident and precise in their mathematics talk.

Effective communication occurs only if both the students and the teachers listen when someone is talking. For the students, they listen patiently and consider the views of the others, wait for their turn to contribute and build on their friends' ideas. They must disagree respectfully, and direct their challenges at the ideas and not the individual students. For the teachers, they listen and orchestrate the mathematics talk in the whole class discussion, listening to students' group discussions, and gathering valuable information for making instructional decisions.

Not all classroom talk is productive. Productive mathematics talk must be accountable to the learning community, to accurate and appropriate knowledge, as well as to rigorous thinking. Hence, what students are talking about and how they talk in class do matter. For talk to be accountable, it would not be spontaneous, but students have to be attentive, accurate and precise in their observations or support their claims by sound reasoning and evidence. The evidence has to be examined critically, accurately and sufficiently (Michaels et al., 2013).

4 Getting Students to Talk in the Mathematics Classroom

It is not easy to get students to talk productively in the mathematics classroom. To organize mathematics talk, teachers must first remember that the main purpose of mathematics talk is to build knowledge and all students must be given opportunity to participate fully in the construction of knowledge through mathematics talk. Research studies have suggested various strategies to promote classroom talk. They include:

- Establishing of recurring, predictable routines with specific talk formats (Michaels et al., 2013).
- Setting up a supportive environment (Clark et al., 2005).
- Use of rich task (Clark et al., 2005).
- Use of talk moves to orchestra the discourse (Chapin, O'Connor, & Anderson, 2003).
- Engaging the art of questioning.

Routines set up by teachers regarding how to participate in mathematics talk in a specific talk format will facilitate classroom discourse. For example, when the students are told to 'partner talk for 5 seconds', they can immediately turn to their partner and discuss the point for the time given. They are aware what topics they are not supposed to talk about and that they are expected to share their view with the class after partner talk. Once students are aware of the routine, they can focus their attention on what is being discussed. For example, ideas have to be explained clearly and disagreements have to be resolved satisfactorily. Or teachers may establish routines when conducting recurring whole class discussions. For example, students are to raise their hands up before they talk, take turns to talk and know how long they are supposed to talk. They learn how to have empathy and direct their disagreement appropriately. Teachers may provide a list of prompts to students who are not familiar with whole class discussion protocol. Such a list provides support to students to engage in thinking during mathematics talk.

A respectful and supportive environment is essential to productive mathematics talk. An eager young student who had something to contribute in the middle of discussion and is told by the teacher to keep quiet and not to interrupt would be reluctant to participate further in class discussion. A shy and sensitive student who finds it difficult to articulate his or her idea in front of the class would need encouragement and prompts. Students would be reluctant to participate in productive mathematics talk if they are silenced or are afraid that they will be laughed at or made to feel stupid. On the other hand, there are also students who tend to dominate class/group discussion while others may just keep quiet even if they have good ideas. A supportive environment helps every student develop their social skills and socializing intelligence. It is a learning environment where students respect and trust each other, are willing to take risks and criticisms, knowing that the comments given by the teacher and their classmates are directed at the ideas and not at them. They believe that it is their right to be heard, obligation to be attentive and participate in the mathematics talk.

Students would participate in productive mathematics talk if there are clear objectives for the lesson and rich instructional tasks to accomplish the objectives. Different tasks will work best with different talk formats. A rich task with high cognitive demand would be able to sustain an extended class or group discussion. Besides considering how to select and launch the task, teachers also have to plan carefully, anticipate what is likely to happen, what directions student talk may take, how to orchestrate the discussion, what support to give and how to achieve the lesson objectives.

The Magic V (NRICH, 2008) shown in Figure 3 is an example of a rich task that allows students to practise addition of three single-digit numbers, apply problem solving skills and develop their number sense and mathematical thinking. It involves big ideas and students can make conjectures and be engaged in complex thinking. Classroom talk carefully orchestrated by teachers will lead students to apply their number sense to find the solutions instead of adopting the trial and error approach.

Place each of the numbers 1 to 5 in the V shape so that the two arms of the V have the same total.

How many different possibilities are there?

Can you convince someone that you have all the solutions?

What happens if we use the numbers from 2 to 6? From 12 to 16? From 37 to 41? From 103 to 107?

Investigate the same problem with a V that has arms of length 4.

Figure 3. The Magic V (NRICH, 2008)

The teacher may begin with the whole class discussion by showing the following four arrangements (Figure 4) and getting the class to compare, identify the variant and the invariant features among the arrangements.

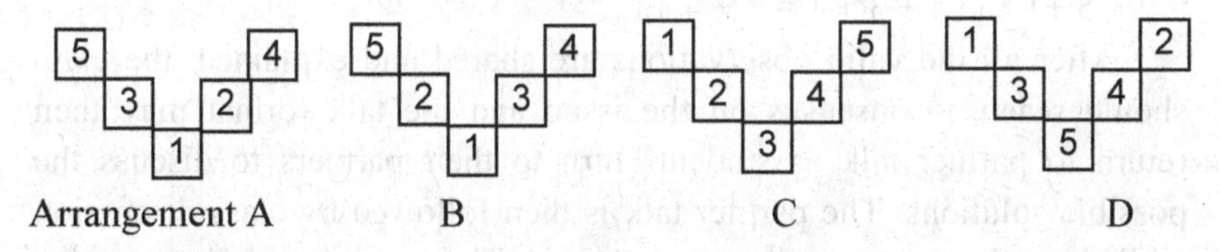

Figure 4. Introduction to the Magic V task

Students may talk about similarities such as "*the same set of 5 numbers 1 to 5 is used in each arrangement*", "*there are 3 odd numbers and two even numbers*", "*the number at the vertex of V is always an odd number.*" They may mention differences such as "*arrangements A and D have the three odd numbers on one side while B and C has an even number on each arm*", "*in arrangements A, C and D, the sum of the numbers at each arm is different but in B, the sums of the numbers at the two arms are the same.*"

The teacher can then revoice the last observation and direct students to 'partner talk' and look for other arrangements that give the same total for each arm of V. As the students work to find the arrangements, question may arise whether the following arrangements (Figure 5) are the same.

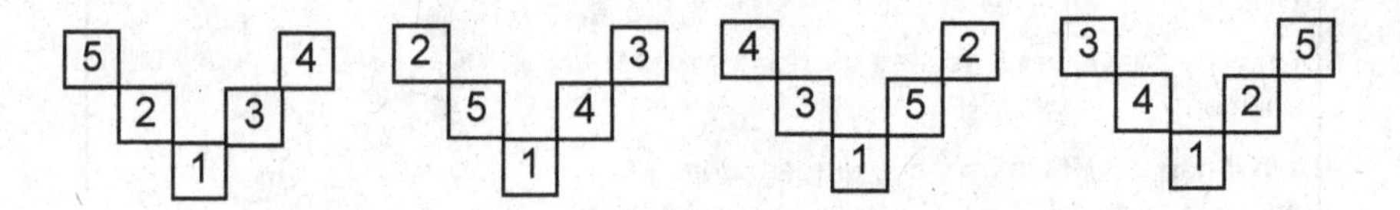

Figure 5. Some possible solutions to Magic V task

The teacher could then interrupt and get the whole class back for discussion. Those students who agree and those who disagree that these arrangements are the same should be encouraged to give reasons to justify their views. It may be useful to provide students sentence starters to guide the class discussion, making sure they use correct and precise language to communicate their thinking. For example, terms like 'commutative property for addition' may be used to justify that all arrangements in Figure 5 could be considered the same because $5 + 2 + 1 = 2 + 5 +1$ and $4 + 3 + 1 = 3 + 4 +1$.

After all the valid observations are shared and explained, the class should reach a consensus on the issue and the talk format may then return to partner talk as students turn to their partners to discuss the possible solutions. The partner talk is then followed by class discussion while hypotheses are made and evaluated. The possible solutions for the task are shown in Figure 6.

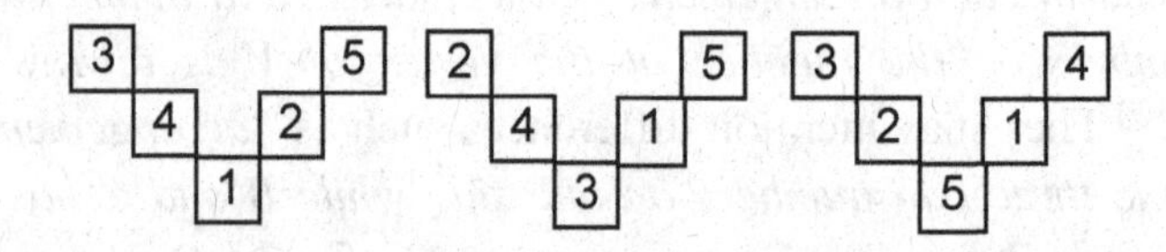

Figure 6. The three solutions to the Magic V task

Students may think that since the smallest number is odd, all the possible solutions would have an odd number at the vertex of V. The teacher can then get students to think by asking:

What happens if we use the digits 5, 6, 7, 8 and 9 instead? Are the strategies used previously still applicable? What happens if the smallest number is even, say we have 2, 3, 4, 5 and 6? What happens if we use 7 consecutive numbers? Can we form a V shape using even number of digits? What happens if we use 8 digits to form a different configuration, like these (see Figure 7)*? How are these two configurations different? How would you best describe them? Why? What is alike and what is different about the method of solution for the Magic V and the solution for these two configurations? Is there a pattern? Explain.*

Questions like these encourage productive mathematics talk. Magic V is a task with clear instructional goals and allows students to exit at different points according to their interest and ability.

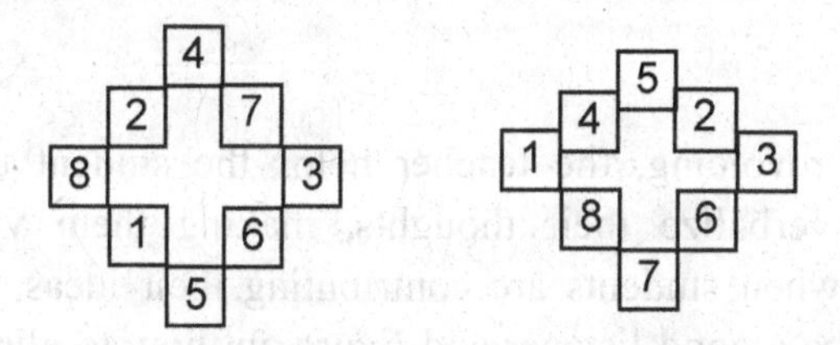

Figure 7. A 'rhombus' and 'kite' configuration

Whichever talk formats the students are engaged in, partner talk, whole class discussion or presentation, teacher can use talk moves to orchestra the discourse. Chapin et al. (2003) identify five moves teachers can use in mathematics talk to help students learn. They are:

- revoicing,
- asking students to restate someone else's reasoning,
- asking students to apply their own reasoning to someone else's reasoning,
- prompting students for further participation, and
- using wait time.

Revoicing involves three basic components. First a student says something, and then the teacher helps students to clarify their own reasoning by repeating what the student has said before asking the student to verify whether the revoicing is correct. Hence, the purpose of the teacher's talk move is to provide more 'thinking space' for other students to follow the mathematical talk in the classroom. The dialogue below illustrates an instance of revoice during the class discussion on the Magic V task.

> *Student: "In arrangement B, 5 and 2 and 1 is 8 ... 4 and 3 and 1 is also 8."*
> *The teacher may proceed to write* $5 + 2 + 1 = 8$ *and* $4 + 3 + 1 = 8$ *on the board and revoice: "Are you saying that the sum of the numbers on the two sides are the same?"*
> *Student: "Yeah....... when we add, we get the sum... both sums are eight."*

Through revoicing, the teacher helps the student use appropriate language to verbalize their thoughts, making them visible to other students. So when students are contributing their ideas, the role of the teacher is to be a good listener and figure out how to align the student's contribution to the instructional goal.

It is another student and not the teacher, who repeats or rephrases what their classmate has said in the talk move "asking students to restate someone else's reasoning". Consequently, the students in the class have another rendition of the first student's response that they may find easier to follow. In this talk move, it is better to ask someone sitting on the opposite side of the classroom to carry out the restating. This move also ensures all students are attentive and can hear what has been said, giving the first contributor the idea that his contribution is valued.

The third move "asking students to apply their own reasoning to someone else's reasoning" is also referred to as "reflective toss" (van Zee & Minstrell, 1997) or "challenging student" (Michaels et al., 2013). Here the teacher is not just asking what one student thinks about the other student's contribution, but is also pressing for reasoning, why a claim is supported or disputed. In the Magic V task, a student may claim that there are only three possible solutions (see Figure 6). The teacher can then ask another student's view about the claim and the reasoning behind the claim.

Classroom participation can be further increased when the teacher asks students for further commentary on the on-going discussion. Here students build on the ideas of their classmates. Consequently, the discussion grows in breadth as well as in depth. For example, in the Magic V task, some students may wonder about the context of the problem change? What happens if

- there are seven instead of five numbers?
- the smallest of the five consecutive numbers is even instead of odd?
- there is an even number of consecutive numbers?
- there are eight numbers arranged in the 'rhombus' or 'kite' configuration, do we still have only three possible solutions?

In order for students to reason rigorously and formulate their response, be it an answer or a question, time is needed. The 3 – 5 seconds of 'wait-time' permits students to think something through, frame a response and communicate their thoughts. Research has shown that increasing 'wait-time' would lead to longer and more rigorous student response and improve the quality of mathematics talk. There are two main kinds of wait-time: Wait-time after the teacher poses a question gives students opportunity to think and generate their response; wait-time after a student has responded gives other students a chance to understand their friend's line of thinking. Many teachers feel uncomfortable when a question is followed by silence. They tend to accept the first answer or call on the first student who raises a hand. There is no time given for the slower students to think about an answer to the question.

The teacher's questions and management of students' responses are as important as the wait-time in promoting mathematics talk. The *Professional Standards for Teaching Mathematics* (NCTM, 1991) identify five types of questions that can be used to stimulate student talk. They include questions that help students

- work together to make sense of mathematics;
- to rely more on themselves to determine whether something is mathematically correct;
- to learn to reason mathematically;
- to learn to conjecture, invent and solve problems; and
- to connect mathematics, its ideas and its application.

The questions posed can be redirecting questions when student responses are unsatisfactory or incomplete, or probing questions to elicit more complete responses from the students and reinforce learning.

5 Conclusion

Mathematics talk is not limited to students answering the teacher's questions and taking turns to show and share their solutions. It is an instructional discourse with the teacher at the helm but with as much student-to-student talk as possible. There are multiple benefits for productive mathematics talk in class. However, building a mathematics talk community that engages all students is a gradual process, and both the teacher and students must have a paradigm shift regarding teaching and learning in the 21st century. The shift is from a teacher-questions students-answer environment to a more vibrant student-student interaction that leads to productive mathematics talk and mathematical knowledge construction. Teachers have to relinquish their roles as the sole questioner and the assessor of learning in class. They should share the role of questioner with the students instead, and have students take greater responsibility for their own learning. They should also explicitly teach the students the acceptable social behaviors necessary for engaging in productive mathematics talk.

References

Chapin, S.H., O'Connor, C., & Anderson, N.C. (2003). *Classroom Discussions: Using Math Talk to Help Students Learn, Grades 1-6.* CA: Math Solutions Publications.

Clark, K.K., Jacobs, J., Pittman, M.E., & Borko, H. (2005). Strategies for building mathematical communication in the middle school classroom: modeled in professional development, implemented in the classroom. *Current Issues in Middle Level Education, 11*(2), 1-12.

Cockcroft Report (1982). *Mathematics Count.* London: Her Majesty's Stationery Office.

Hogan, D. (2014). *Why is Singapore's school system so successful, and is it a model for the west?* The Conversation. Retrieved 10 December, 2015 from http://theconversation.com/why-is-singapores-school-system-so-successful-and-is-it-a-model-for-the-west-22917

Hogan, D., Rahim, R.A., Chan, M., Kwek, D., & Towndrow P. (2012). Understanding classroom talk in secondary three mathematics classes in Singapore. In B. Kaur, & T.L. Toh (Eds.), *Reasoning, Communication and Connections in Mathematics,* (pp. 169-197). Singapore: World Scientific.

Ministry of Education, Singapore (2010). *MOE to enhance learning of 21st century competencies and strengthen art, music and physical education.* Retrieved 31 December, 2015 from www.moe.gov.sg

Ministry of Education (2012). *Primary Mathematics: Teaching and Learning Syllabus.* Singapore: Curriculum Planning and Development Division.

Michaels, S., O'Connor, M.C., Hall, M.W., & Resnick, L.B. (2013). *Accountable talk sourcebook for classroom conversation that works.* University of Pittsburg.

National Council of Teachers of Mathematics (1991). *Professional standards for teaching mathematics.* Reston, VA: Author.

National Council of Teachers of Mathematics (2000). *Principles and standards for school mathematics.* Reston, VA: Author.

NRICH. (2008). *The Magic V.* Retrieved 1 June, 2015 from http://nrich.maths.org/content/id/6814/NRICH-poster_MagicV.png

Seefeldt, C. (2004). Helping children communicate. Early Childhood Today. Scholastic, Retrieved 15 May, 2015 from http://www2.scholastic.com

van Zee, E.H., & Minstrell, J.A. (1997). Reflective discourse: Developing shared understanding in a physics classroom. *International Journal of Science Education, 19*, 209-228.

Wagganer, E.L. (2015). Creating math Talk Communities. *Teaching Children Mathematics, 22*(4), 248-254.

References

Chapter 10

Justification in Singapore Secondary Mathematics

CHUA Boon Liang

Justification is a key process skill that students require in the learning of mathematics. Researchers in the mathematics education community have always shown an interest in how students reason and express justifications. With growing interest in mathematical reasoning and justification beyond mathematical proof, the importance of justification tasks therefore grows. This chapter begins with a discussion of justification and its importance, followed by a description of the types of justification tasks that mathematics students are commonly asked to do as well as the demand of each type of task. The chapter continues by examining the performance of 22 Grade 9 Singapore students and 50 mathematics teachers in two justification tasks. Students appear to struggle with the tasks whilst teachers seem capable of doing such tasks although many did not cite the right reason in one of the tasks. When asked to score student justifications, the majority of teachers were accurate in their assessment, indicating they were capable of recognising an acceptable justification. Teaching strategies for promoting mathematical justification in the classroom based on these findings are then suggested.

1 Introduction

Given the emphasis on 21st century competencies (21cc), greater demands are being placed on students to reason, explain and justify in the learning of mathematics. Mathematical reasoning refers to the ability to analyse mathematical situations and construct logical arguments (Ministry of Education, Singapore, 2012). The articulation of the mathematical ideas or arguments to explain a mathematical situation or to convince others of its validity is part of what is called mathematical justification. Clearly, both mathematical reasoning and justification are essential and inseparable components of any mathematical activity. Reasoning as a didactic device in the learning of mathematics cannot avoid some degree of justification because getting students to justify is one powerful means to gain an insightful perception of their thinking and reasoning. When students are able to analyse, reason and communicate ideas clearly as they perform the mathematical activity, they are considered mathematically literate. Mathematical literacy is crucial for the development of two 21cc as identified by the Ministry of Education, Singapore (MOE): *critical thinking* and *communication skills* (Ministry of Education, Singapore, 2010).

Mathematical reasoning and justification are, however, not new process skills. For years, they have been included under the *Processes* component in the Singapore Mathematics Framework. Amongst the many key processes highlighted in this component are reasoning, communication and connections (Ministry of Education, Singapore, 2012). Communication in mathematics refers to the ability to use the language of mathematics to express mathematical ideas and arguments precisely, concisely and logically. Mathematical justification, which is described in greater detail in the next section, thus falls under communication in the framework. So mathematical reasoning and justification should have been encouraged in all mathematics lessons in Singapore although the extent of them being practised is not clear and may vary across different classes. Further, not only do secondary school mathematics teachers seem to perceive justification tasks in GCE O level and N level examinations as higher-order thinking questions meant for the more able students, students also tend to dodge such questions in tests and examinations. The teachers' perceptions of justification tasks, together with the students' attitude towards such tasks, therefore make mathematical justification even more worthwhile to investigate. This chapter therefore sets out to examine (i) the way Singapore secondary school students and mathematics teachers express justifications, and (ii) mathematics teachers' assessment of the students' justifications. It is structured along the following strands of work:

- a perspective of what justification is and why it is important for the learning of mathematics,
- a view of justification tasks and what they demand from students,
- an overview of the responses produced by a group of Singapore secondary school students and mathematics teachers to two justification tasks and of how mathematics teachers have assessed the students' justifications.

2 What is Justification?

The topic of justification is frequently linked to the topic of proof in the literature. According to Simon and Blume (1996), mathematical justification is the process of "establishing validity [and] developing an argument that builds from the community's taken-as-shared knowledge" (p. 28). The notion of justification as a means of determining and explaining the truth of a mathematical conjecture or assertion resonates strongly with many other researchers. For instance, Balacheff (1988) described justification as "the basis of the validation of the conjecture" (Balacheff, 1988, p. 225) – a view supported by Huang (2005) and Thomas (1997) as well. To Harel and Sowder (1998, 2007), validation as one form of justification was not just about *ascertaining* the truth of the conjecture or assertion, but also about *persuading* others whether or not it is true. Whilst the process of ascertaining the truth involves removing one's own doubts, the process of persuading is one's attempt to remove others' doubts (Ellis, 2007). What both processes of validation do share in common is the role of conviction.

The types of responses expected of students in the justification process depend on at least two factors: *the cognitive abilities of students* and *the nature of the task*. For secondary school students, particularly those in the lower grades, a justification does not need to measure up to a formal proof. This is because providing a theoretical argument for a mathematical result is sometimes not required in the light of their cognitive level until they reach higher level of study (Hoyles & Healy, 1999). Take, for instance, the justification task asking lower secondary school students why $2n - 1$ is an odd number for any positive integer n. An acceptable justification could simply state: with n being any positive integer, forming two groups of n, which can be expressed as $2n$ in notation, thus generates an even number, therefore subtracting one from it will result in an odd number.

Not all justification tasks require a theoretical argument, however. Some lend themselves well to experiential justification, which is mainly supported by specific examples and illustrations. Consider asking students to justify why the rule $a^m \times a^n = a^{m+n}$ is true for any positive integer *a*, *m* and *n*. The students rely on intuitive reasoning using several numerical examples in the justification: for instance, $2^3 \times 2^4 = (2 \times 2 \times 2) \times (2 \times 2 \times 2 \times 2) = 2^7 = 2^{3+4}$, $3^2 \times 3^5 = (3 \times 3) \times (3 \times 3 \times 3 \times 3 \times 3) = 3^7 = 3^{2+5}$, and $4^5 \times 4^3 = (4 \times 4 \times 4 \times 4 \times 4) \times (4 \times 4 \times 4) = 4^8 = 4^{5+3}$, so the rule works for any positive integer *a*, *m* and *n*. What students are usually expected to do when asked to justify a mathematical result is to explain to others why it is true. Such a justification does not involve any theorems and is therefore deemed a less formal argument than a typical mathematical deductive proof (Becker & Rivera, 2009). But it is this type of justification that is valued because it "explains rather than simply convinces" (Lannin, 2005, p. 235). Unlike a proof-related justification which is normally expected to be presented in the written form, the explanatory type can take the form of an oral justification as well.

A justification does not always have to be an explanation of why a mathematical result is true. As Becker and Rivera (2009) pointed out, a justification can be an elaboration of how a mathematical result is obtained. In the topic of pattern generalisation, for instance, students can be asked to show clearly how they established a rule for the pattern. What they have to do then is to illuminate the method used to construct the rule.

To sum up, justification is the process of making a statement in response to a few different types of requests: (1) *conviction*, which is concerned with establishing the truth of a mathematical result, (2) *explanation*, which is concerned with conveying insight into why a mathematical result is true, and (3) *elaboration*, which is concerned with illustrating the method used to obtain a mathematical result. The justification can be verbalised or written down, and formal or informal.

3 Importance of Justification

What benefits does engaging in mathematical justification bring to learning? First and foremost, Mathematics as a discipline calls for students to be able to examine and evaluate the validity of facts, articulate their reasons for employing a certain method to solve a mathematical task, and substantiate any arguments put forth. So justification, alongside mathematical reasoning, is a crucial process skill that enables students to carry out all those activities. Through the activities, students acquire and apply mathematical knowledge and skills.

The mathematical knowledge that students acquire enables them to understand and make sense of the world because mathematical understanding lays the foundation of economic, scientific and technological progress. With mathematics all around us and underpinning much of our daily lives, it is therefore of fundamental importance to ensure that the students have good grounding in mathematics and relevant skills to evaluate and justify mathematical situations that they encounter in life in the future.

Justification helps to scaffold students' thinking and reasoning when teachers guide students to analyse, interpret and critique their own work, as well as those of others in class. As the students express justification, any misconceptions that they may have and their lines of thoughts become visible. Probing students' understanding and encouraging them to justify their thinking also lead to a deepening of their understanding of mathematical concepts and processes. As well, the students are engaged in active construction of mathematical knowledge when they internalise and consolidate their thinking and understanding of concepts.

Justification is also a useful tool for helping students to develop and hone their critical thinking skills, which are highlighted as a key competency in the 21cc framework developed recently by MOE (Ministry of Education, Singapore, 2010). Critical thinking skills entail considering mathematical situations from various perspectives, evaluating and analysing the strengths and weaknesses of the different perspectives, and justifying the validity of the mathematical situations. Engaging students actively in justification pushes them to be discerning in judgment, helps to sharpen their critical thinking skills, and promotes effective and precise communication of ideas. All of these are critical attributes of a *confident* person in the 21st century that MOE aims to develop in students (Ministry of Education. Singapore, 2010).

Presenting mathematical justifications orally or in writing involves processes that are fundamental to learning. As students attempt to explain mathematical concepts and justify procedures, they need to clarify, organise and consolidate their thinking and understanding of concepts and procedures. For effective communication to take place, not only do the students have to organise their ideas in a systematic and logical way, equally important, they also have to articulate them in a precise manner as well. As pointed out previously, communication skills are identified as one of the crucial 21cc because effective and precise communication allows the receiver to fully understand what the speaker or writer has conveyed. This then reduces any ambiguity and misunderstanding, thereby enabling people to work effectively in the global landscape of the 21st century. Thus student participation in justification allows them to cultivate and develop those crucial processes and should therefore be encouraged in mathematics teaching.

4 Justification Tasks

Two roles of justification are gleaned from the literature review: *elaboration* and *validation*. Thus justification tasks can be described as questions asking students to elaborate, explain, and validate a mathematical result. These tasks are integral to mathematics teaching and are prevalent in the GCE O level and N level examinations. Figure 1 below presents an example of a justification task on the topic of number patterns that illustrates those two roles of justification.

> The first five terms of a sequence are given below. Each term in this sequence is found by adding the same number to the previous term.
>
> 5, 12, 19, 26, 33, …
>
> (a) Peter produced the expression $33 + 7(n - 5)$ for the nth term of the sequence. Explain how he might have obtained this expression.
>
> (b) Explain why 139 is not a term of this sequence.

Figure 1. A justification task on number patterns

The justification task in Figure 1 comprises two part questions. Part (a) provides an algebraic expression $33 + 7(n - 5)$ and requires students to elaborate how it is derived. In part (b), the term 139 is given and made known to the students that it does not belong to the sequence. The students are then required to provide supporting evidence to show why that is so.

The nature of justification tasks is however not necessarily limited to asking students to validate and elaborate a mathematical result. Two other types of questions will be illustrated using the number pattern task in Figure 1. The first type involves making a prediction with inference from facts: for instance, explain whether 110 is a term of the sequence. Although this task and part (b) in Figure 1 look alike, there is a subtle distinction between them in terms of the task nature. In the present task, it is not clear whether 110 follows the same number pattern, unlike the situation in part (b). So students will have to decide and then draw a conclusion with supporting evidence to say if 110 belongs to the same number sequence formed by 5, 12, 19, 26 and 33. The other type asks for an interpretation of a mathematical result: for instance, explain what the coefficient of n in $7n - 2$ represents in the sequence. Such a task expects students to give the significance of the object in question: that is, the coefficient of n in the example.

To sum up, justification tasks can be classified into four categories: *elaboration*, *interpretation*, *prediction* and *validation*. With different nature of the task in each category, the demand of each type of task is therefore not the same. Table 1 provides an overview of the four types of justification tasks and what each type requires students to do.

Table 1
Nature and demand of justification tasks

Nature of justification tasks	Examples	What students are expected to do
Elaboration	Explain how…	Describe / show clearly the method or strategy used to obtain the mathematical result
Interpretation	Explain what…	Give the meaning of the mathematical result
Prediction	Explain whether… Explain which…	Make decision about the mathematical claim and provide evidence to support or refute the claim
Validation	Explain why…	Give reason or evidence to support the mathematical claim

Examples are now provided to illustrate each of the four types of justification tasks. Part (a) in Figure 1 above is an *Elaboration* task on number patterns. *A Stone's Throw* in Figure 2a is an algebra task belonging to the *Interpretation* type of justification task.

A stone was thrown from the top of a vertical tower. Its position during the flight is represented by the equation

$$y = 50 + 21x - x^2,$$

where y metres is the height of the stone above the ground and x metres is its horizontal distance from the foot of the tower.
Explain what the positive solution of the equation $0 = 50 + 21x - x^2$ represents.

Figure 2a. A Stone's Throw task

Mr. Right Triangle in Figure 2b is a geometry task of the *Validation* type whilst *Cybertime* task in Figure 2c belongs to the *Prediction* type.

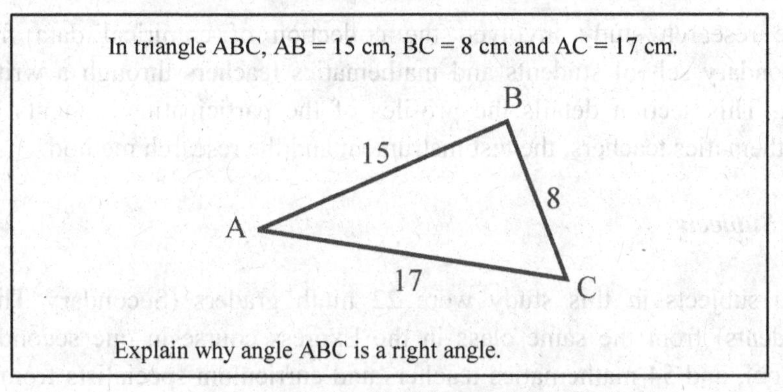

Figure 2b. Mr. Right Triangle task

The time (in hours) that 18 children spent on the internet in a certain week are shown below.

21	24	26	68	22	19
23	25	21	23	25	21
51	22	23	25	26	21

Explain which of the three averages is the <u>most suitable</u> for these data.

Figure 2c. Cybertime task

The remaining section reports on (i) secondary school students and mathematics teachers' justifications to two justification tasks which formed part of a written test developed for the current project, and (ii) the teachers' assessment of the students' justifications. The research questions that are addressed in this chapter are provided below:

(1) How do secondary school students in Singapore perform in justification tasks?
(2) How do mathematics teachers in Singapore perform in justification tasks?
(3) How do mathematics teachers in Singapore assess students' justifications?

5 A Research Study on Mathematical Justification

The research study involved the collection of empirical data from secondary school students and mathematics teachers through a written test. This section details the profiles of the participating students and mathematics teachers, the test instrument and the research method.

5.1 *Subjects*

The subjects in this study were 22 ninth graders (Secondary Three students) from the same class in the Express course in one secondary school, and 54 mathematics teachers and curriculum specialists from 32 different secondary schools and MOE. Both the student and teacher samples were obtained from convenience sampling. There were 14 girls and eight boys in the student sample. These students were 15 years old and did not take Additional Mathematics, a more challenging O-level subject, as they were considered to be academically less able in the school's Year 9 cohort. Regarding the teachers, they attended an in-service workshop on mathematical justification conducted by the researcher in June 2015. Thirty percent of them taught mathematics for less than 5 years, another 40% for at least five to less than 15 years, and the remaining 30% for 15 years or more.

5.2 *Test instrument*

A paper-and-pencil test was designed to assess students and mathematics teachers' ability to justify. The student version of the test consists of five justification tasks covering the following strands in the Singapore mathematics curriculum: Number, Algebra, Geometry and Statistics. Of the five tasks, two were algebra tasks and there was one for each of the remaining three strands. These tasks were adapted from past year questions in the GCE O level and N level examinations. Each task carries one mark, following the same amount normally allocated in the GCE examinations. The teacher version had the same justification tasks, except that one of the two algebra tasks was dropped and that the test now contained a second section that was designed to investigate how the teachers assess students' justifications. This chapter discusses the students and teachers' responses to only two of the tasks, *A Stone's Throw* (see Figure 2a) and *Mr. Right Triangle* (see Figure 2b).

The first section of the test in the teacher version served not only to examine the teachers' justifications to the four tasks but also to familiarise them with these tasks to better prepare them to understand the tasks that they had to do later in the second section. The second section comprised the four justification tasks that the teachers had earlier attempted, each accompanied by two authentic student solutions selected from the ninth graders' responses. With each task worth only a mark, the teachers had to score each student solution, awarding either one mark for a correct response or zero for an incorrect response. Figures 3a and 3b below show the two student solutions for *A Stone's Throw* and for *Mr. Right Triangle* respectively.

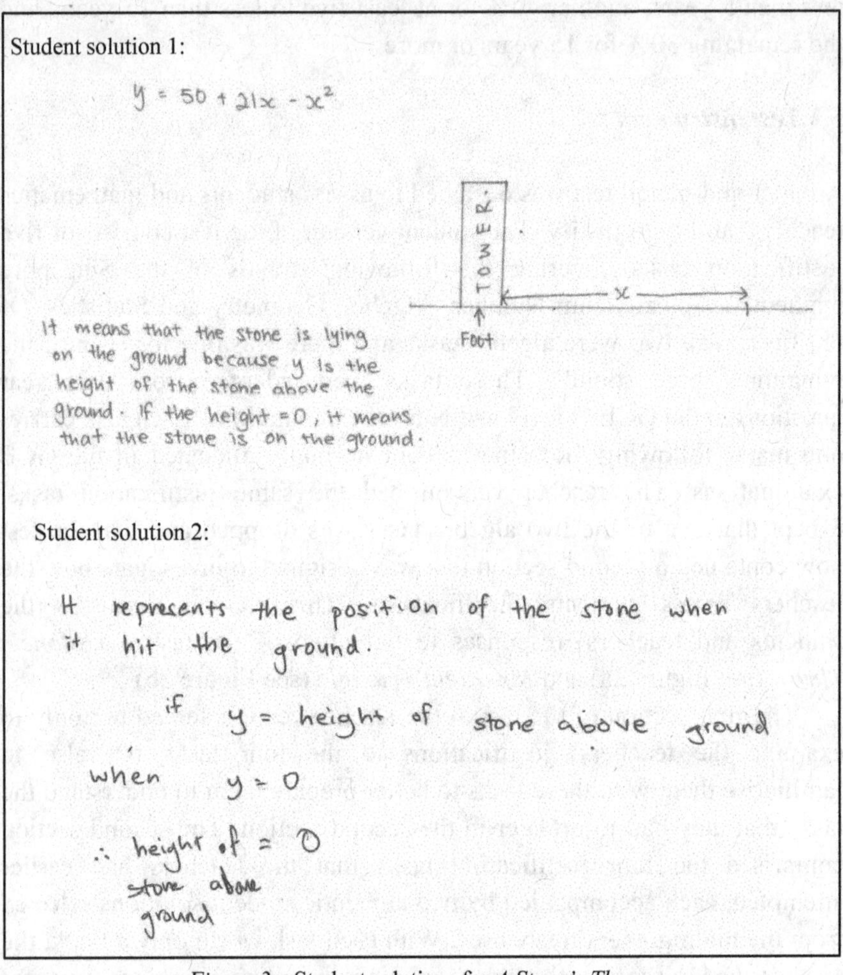

Figure 3a. Student solutions for *A Stone's Throw*

5.3 Procedures

This study used a survey design to gather quantitative data through a written test administered to 22 ninth graders in April 2015 and to 54 mathematics teachers in June 2015 during a two-hour in-service workshop. The students were given 45 minutes to complete all five justification tasks in the test whereas the teachers had to complete all four justification tasks and the marking of eight student solutions in 30 minutes. Prior to participating in this study, the students had learnt all the concepts tested in the test.

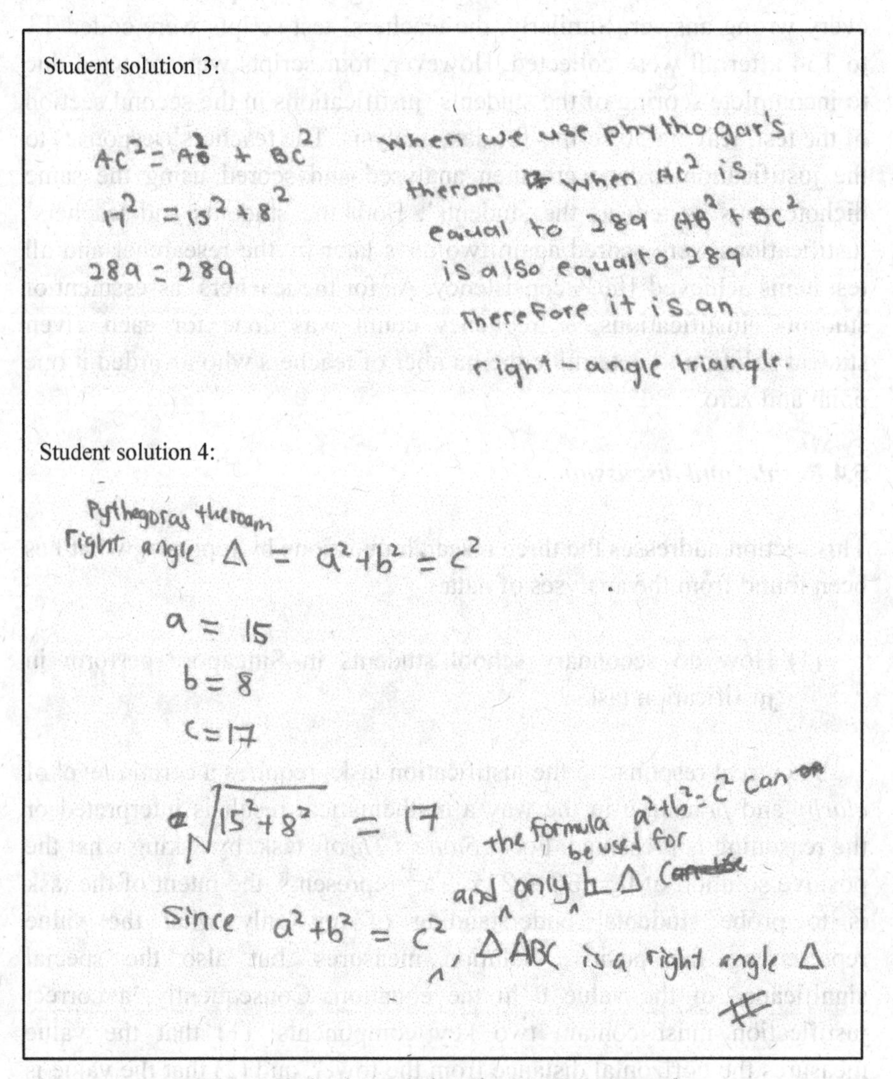

Figure 3b. Student solutions for *Mr. Right Triangle*

All the students' test scripts were collected and immediately, in the order they were collected, coded F1 to F14 and M1 to M8, where F denotes female student and M denotes male student. Subsequently, all the students' responses were analysed carefully and then scored dichotomously, with every correct answer given one point and zero for every wrong answer. Similarly, the teachers' test scripts were coded T1 to T54 after all were collected. However, four scripts were rejected due to incomplete scoring of the students' justifications in the second section of the test, leaving 50 scripts for data analysis. The teachers' responses to the justification tasks were then analysed and scored using the same dichotomous system as the students'. Both the students and teachers' justifications were scored again two days later by the researcher and all test items achieved 100% consistency. As for the teachers' assessment of students' justifications, a frequency count was done for each given student solution to determine the number of teachers who awarded it one point and zero.

5.4 *Results and discussion*

This section addresses the three research questions by reporting what has been found from the analyses of data.

(1) How do secondary school students in Singapore perform in justification tasks?

A correct response to the justification tasks requires a certain level of *clarity* and *precision* in the way a mathematical result is interpreted or the reasoning is presented. For *A Stone's Throw* task, by asking what the positive solution of $0 = 50 + 21x - x^2$ represents, the intent of the task is to probe students' understanding of not only what the value representing the positive solution measures, but also the special significance of the value 0 in the equation. Consequently, a correct justification must contain two key components: (1) that the value measures the horizontal distance from the tower, and (2) that the value is measured at the instance when the stone hit the ground. For *Mr. Right Triangle* task, a correct justification must check that the condition $8^2 + 15^2 = 17^2$ is satisfied followed by citing the converse of Pythagoras theorem as a warrant for drawing the conclusion that angle ABC is a right angle.

Table 2 shows the frequency of correct answer by students and teachers for each of the two justification tasks, and the frequency of each score obtained by the students and teachers for the two tasks combined. Clearly, two students answered *A Stone's Throw* correctly and none for *Mr. Right Triangle*.

Table 2
Performance of students and teachers in two justification tasks

	A Stone's Throw	Mr. Right Triangle	Total score for both tasks		
	Correct (%)	Correct (%)	0	1	2
Students (n = 22)	2 (9%)	0 (0%)	20 (91%)	2 (9%)	-
Teachers (n = 50)	42 (84%)	14 (28%)	6 (12%)	34 (68%)	10 (20%)

In *A Stone's Throw*, the response "the horizontal distance of the stone from the foot of the tower" occurred twice and was initially considered as incorrect because it was a mere repetition of the definition[1] of x described in the question. After much deliberation, this response was accepted because the phrase "from the foot of the tower" pointed towards the fact that the stone had hit the ground. Incorrect student responses include the following: the stone is lying on the ground (e.g., see Student solution 1 in Figure 3a), the horizontal height of the stone above ground, the position of the stone after landing (e.g., see Student solution 2 in Figure 3a), the area that the stone will land, and the gradient of the equation. Student solution 1 was considered wrong despite being

[1] The definition of x was taken from the original O-level question. On reflection, it would have been better to define x as "the horizontal distance from the tower".

accompanied by a somewhat reasonably labelled diagram because the written part did not correspond with the diagram and capture clearly what the positive value measures. By describing the stone as lying on the ground, the response showed the circumstance the stone is in instead of interpreting the measurement represented by the value. Likewise in Student solution 2, it is linguistically incorrect to refer to the positive value as "the position of the stone". By "position", the response focused on a static point rather than the distance between two points which is what the value represents. The appearance of these erroneous responses indicates that many students did not appreciate the significance of the positive solution of the given equation. Only a small number of students had the right idea but failed to express it precisely.

In *Mr. Right Triangle*, five students nearly produced a correct response. They checked that $8^2 + 15^2$ and 17^2 are equal but provided either no warrant or the wrong warrant – Pythagoras' theorem instead of its converse (e.g., see Student solutions 3 and 4 in Figure 3b). Their responses thus did not score.

Another five students assumed Pythagoras' theorem to be true at the very outset and immediately established $8^2 + 15^2 = 17^2$ without verifying its validity (e.g., see Solution by M4 in Figure 4). Such responses also did not gain the mark. Further, a few students even drew the conclusion simply by visual inspection of the angle ABC in the given figure (e.g., see Solution by F6 in Figure 4). The occurrence of these unacceptable responses indicates that many students did not understand what they were required to produce in their justifications when deducing their conclusions.

(2) How do mathematics teachers in Singapore perform in justification tasks?

Table 2 indicates that a significant majority of mathematics teachers were successful in answering *A Stone's Throw* with a success rate of 84%, but in *Mr. Right Triangle*, the percentage dipped to 28%. Over four-fifths of the teachers (88%) answered at least one task correctly.

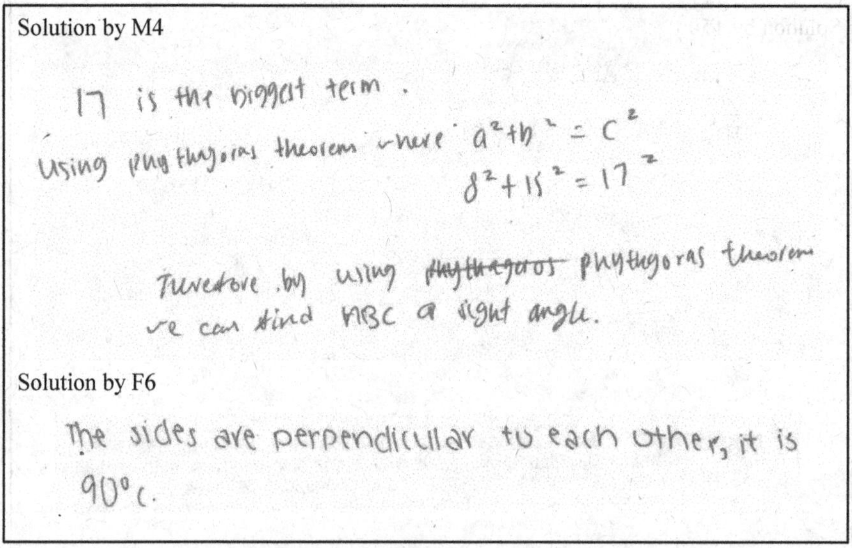

Solution by M4

17 is the biggest term.

Using phythagoras theorem where $a^2+b^2=c^2$

$8^2+15^2=17^2$

Therefore by using ~~phythagoras~~ phythagoras theorem we can find ABC a right angle.

Solution by F6

The sides are perpendicular to each other, it is 90°c.

Figure 4. Incorrect student solutions for *Mr. Right Triangle*

A Stone's Throw was straightforward for the mathematics teachers given its familiar context, with only eight teachers (16%) failing to gain the mark here. The erroneous responses produced by the eight unsuccessful teachers include: when the stone is next at ground level, the maximum height of the throw, the stone reached the ground, and time taken for the stone to touch the ground. The evidence suggests that the majority of mathematics teachers recognised what the *positive* solution of the equation $0 = 50 + 21x - x^2$ represented and were able to express it clearly and precisely.

Mr. Right Triangle proved more challenging and only a small proportion of mathematics teachers were successful in giving a clear and complete justification. This finding was not surprising because personal interactions with other in-service teachers not involved in this study have revealed their inexperience with the justification approach in questions of this type. However, on closer examination of the unsuccessful teachers' responses, it was discovered that of the 36 incorrect responses, 24 of them showed the wrong warrant (e.g., see Solution by T50 in Figure 5).

Solution by T50

Explain why angle ABC is a right angle.

$AC^2 = 17^2$ $BC^2 = 8^2$ $AB^2 = 15^2$
$= 289$ $= 64$ $= 225$

$AC^2 = 289$

$BC^2 + AB^2 = 64 + 225$
$= 289$

$\therefore AC^2 = BC^2 + AB^2$

By pythagoras' theorem, ΔABC is a right $\angle$ Δ. Since AC is the longest length: $\therefore \angle ABC$ must be ~~the~~ 90°

3

Solution by T47

$8^2 + 15^2 = 64 + 225$
$= 289$
$= 17^2$

Hence $\angle ABC$ is a right angle

Solution by T39

$15^2 + 8^2 = 225 + 64 = 289$
$17^2 = 289$
$15^2 + 8^2 = 17^2$ (converse of Pythagora' Thm)

Figure 5. Incorrect teacher solutions for *Mr Right Triangle*

Four were regarded insufficient without any warrant being cited as supporting evidence (e.g., see Solution by T47 in Figure 5), and another three were incomplete without drawing any conclusion (e.g., see Solution by T39 in Figure 5). These results indicate that a significant number of teachers did understand what was required to deduce their conclusions: that is, to check that the condition $8^2 + 15^2 = 17^2$ holds before substantiating the claim with the *converse* of Pythagoras' theorem. But the compelling evidence of nearly half of the teachers citing the wrong warrant in their responses pointed to a possible misconception amongst them: that the Pythagoras' theorem and its converse might have been regarded as essentially the same – an observation similarly noted by Wong (2015).

(3) How do mathematics teachers in Singapore assess students' justifications?

The participating mathematics teachers were asked to individually score authentic student solutions using a dichotomous scoring scale with 1 point for a correct response and zero for an incorrect response. The distribution of their scores for each of the four student solutions related to the two justification tasks discussed in this chapter is presented in Table 3.

Table 3

Distribution of teachers' scores for each student solution

	A Stone's Throw				Mr. Right Triangle			
	SS 1		SS 2		SS 3		SS 4	
Score	0	1	0	1	0	1	0	1
Number of teachers	40	10	38	12	31	19	21	29

SS: Student solution

In *A Stone's Throw*, the distributions of teachers' scores were consistent for both Student solutions 1 and 2, with 80% of the teachers judging the solutions as incorrect. This result corresponds well with the finding above that many mathematics teachers understood what the positive solution of $0 = 50 + 21x - x^2$ referred to. This is why a large number of teachers were able to spot the mistake in each student solution.

On the other hand, there was variation in the distribution of teachers' scores between Student solutions 3 and 4 in *Mr. Right Triangle*. Sixty percent of the teachers gave zero to Student solution 3 but the corresponding percentage of teachers for Student solution 4 dropped to 40%. The analysis indicates that a considerable number of teachers were capable of recognising that Student solution 3, despite the correct validation of the condition, was clearly flawed because the wrong warrant was cited. When the warrant was not provided in a similar solution such as Student solution 4, the solution received one mark from more teachers. This finding demonstrates on one hand that the teachers are more lenient and likely to accept the response as correct when it is not, but on the other hand, it uncovers the possibility that the teachers are not familiar with the rigour of justification of this nature.

6 Strategies for Promoting Justification

Several practical teaching strategies for teachers who want to promote mathematical justification in the classroom emerge from the present study.

Strategy 1.
Introduce the different types of justification tasks available to students. Make clear to them what each type is asking for and what they are expected to produce for their justifications.

Strategy 2.
Guide students to carry out a task analysis to identify what needs to be done for the justification task. To illustrate an example, consider *Mr. Right Triangle*. Teachers can ask the following questions:

- Circle the key word that tells you what you need to do.
- What does this word instruct you to do?

- Underline the mathematical claim that you need to substantiate.
- Establish the givens: What information about triangle ABC are you given?
- Establish the goal: What do you need to show?
- What mathematical concept/theorem relates the givens and the goal?
- Which one do we apply: Pythagoras' theorem or the converse of Pythagoras' theorem?
- (after the justification is completed) Can you substantiate the claim in a different way?

Strategy 3.
Show students good examples of what a clear and precise justification to a task will look like and explain how it addresses the specific demands of the task. A clear and precise justification includes all steps and relevant information organised in a logical manner. Use diagrams to present ideas clearly where possible, and accompany them with a brief written description. Non-examples can sometimes be used for illustration. Consider, for instance, the response "distance from the foot of the tower" for *A Stone's Throw*. Point out that this response lacks precision because it does not suggest the stone has hit the ground. Moreover, it lacks clarity because it does not specify the direction in which the distance is measured and the object whose distance is being measured.

Strategy 4.
Encourage students to make their verbal or written justifications simple and clear using correct mathematical language and symbols. To foster precise communication, advise them to keep their sentences short.

Strategy 5.
Select different students' responses for classroom discussion to highlight what is acceptable and what needs to be revised. The use of authentic student responses convinces students that the errors are not fabricated.

7 Conclusion

The justification process appears to be fraught with difficulties, with many students and even some mathematics teachers failing to navigate this process successfully. Getting students to construct *good justification* takes time especially when they have not much experience doing it in class. It is therefore important that teachers remain patient and continue to probe their students for justifications. Given time and sufficient practice, students will become more confident in articulating their thinking and reasoning, and be able to produce good justifications.

References

Balacheff, N. (1988). Aspects of proof in pupils' practice of school mathematics. In D. Pimm (Ed.), *Mathematics, Teachers and Children* (pp. 216 - 230). London: Hodder and Stoughton.

Becker, J. R., & Rivera, F. D. (2009). Justification schemes of middle school students in the context of generalisation of linear patterns. In M. Tzekaki, M. Kaldrimidou, & H. Sakonidis (Eds.), *Proceedings of the 33rd Conference of the International Group for the Psychology of Mathematics Education* (Vol. 5, pp. 9-16). Thessaloniki, Greece: PME.

Ellis, A. B. (2007). Connections between generalising and justifying: Students' reasoning with linear relationships. *Journal for Research in Mathematics Education*, *38*(3), 194 - 229.

Harel, G., & Sowder, L. (1998). Students' proof schemes. In E. Dubinsky, A. Schoenfeld, & J. Kaput (Eds.), *Research on collegiate mathematics education* (Vol. III, pp. 234 - 283). Providence, RI: American Mathematical Society.

Harel, G., & Sowder, L. (2007). Towards comprehensive perspectives on the learning and teaching of proof. In F. K. Lester, Jr. (Ed.), *Second handbook of research on mathematics teaching and learning* (pp. 805 - 842). Charlotte, NC: Information Age Publishing.

Hoyles, C., & Healy, L. (1999). Students' views of proof. *Mathematics in School*, May, 19 - 21.

Huang, R. (2005). Verification or proof: Justification of Pythagoras' theorem in Chinese mathematics classrooms. In H. L. Chick, & J. L. Vincent (Eds.), *Proceedings of the 29th Conference of the International Group for the Psychology of Mathematics Education* (Vol. 3, pp. 161-168). Melbourne, Australia.

Lannin, J. K. (2005). Generalising and justification: The challenge of introducing algebraic reasoning through patterning activities. *Mathematics Thinking and Learning*, *7*(3), 231 - 258.

Ministry of Education, Singapore (2010). *MOE to enhance learning of 21st century competencies and strengthen art, music and physical education*. Retrieved 2 November, 2015 from www.moe.gov.sg

Ministry of Education, (Singapore. (2012). *O-Level Mathematics Teaching and Learning Syllabus*. Singapore: Curriculum Planning and Development Division.

Simon, M. A., & Blume, G. W. (1996). Justification in the mathematics classroom: A study of prospective elementary teachers. *Journal of Mathematical Behaviour*, *15*, 3 - 31.

Thomas, S. N. (1997). *Practical reasoning in natural language*. (4th ed.). Upper Saddle River, NJ: Prentice-Hall.

Wong, K. Y. (2015). *Effective mathematics lessons through an eclectic Singapore approach*. Singapore: World Scientific.

Chapter 11

Examples in the Teaching of Mathematics: Teachers' Perceptions

Lay Keow NG Jaguthsing DINDYAL

As part of a study examining how teachers in Singapore select and use mathematical examples, 121 teachers from 24 secondary schools responded to three open-ended questions about the use of examples in the teaching of Mathematics. This paper reports findings from a questionnaire that was used in the selection of teacher participants for the main research. All the teachers involved have at least five consecutive years of mathematics teaching experience, with a mean of 12 years, and have had some experience teaching at the upper secondary level. The results showed that students' abilities and the difficulty level of the examples were among the topmost considerations teachers have when introducing mathematical ideas or when selecting homework tasks. There were also noticeable variations in teachers' example choice for different instructional purposes. In addition, this paper also reports on teachers' perceptions of a "good" example and revealed a connection between teachers' knowledge and their examples.

1 Introduction

In line with nurturing 21st century competencies, the Ministry of Education (MOE) has proposed the idea of "Enabling Teachers" so that amongst others, teachers will have more opportunities to develop themselves professionally (see http://moe.gov.sg). For mathematics teachers, one way to achieve this goal is to identify what the teachers value when they make important decisions in selecting examples for instructional purposes, following which, appropriate professional development (PD) courses can be organized for them. The use of examples by teachers in the teaching of mathematics is a well-established practice. Examples are omnipresent not only in the modern-day classrooms but are also discernible in historical records like the Egyptian papyri or the Babylonian tablets. Despite the routine and widespread use of examples for instruction, the selecting and crafting of examples involves the contemplation of many complex and often conflicting factors, which sometimes have to be decided by the teachers on the spot. Thereafter, the enactment of the chosen examples requires careful orchestration and tailoring in order to accurately shape students' understanding and to reduce ambiguity to the least.

While researchers have attended to the roles of sub-categories of examples, research into how teachers integrate examples into their teaching remains scarce (Zodik & Zaslavsky, 2008). Related studies have only begun about ten years back at different parts of the world and till date, it remains unexplored in Singapore. Research has also shown that the use of examples or exemplification in short, is neither arbitrary nor straightforward, where both prospective teachers (Huntley, 2013) and experienced teachers (Zodik & Zaslavsky, 2008) face problems, hence summoning the need for research in this area.

Literature has also revealed a strong connection between teachers' knowledge and their provision of examples in teaching. Rowland, Huckstep and Thwaites (2005) found that teachers' ability in selecting suitable mathematical examples was strongly related to their mathematics content knowledge for teaching. Also, Chick (2010) stressed that the capacity of teachers in crafting effective examples relies heavily on their pedagogical content knowledge too. Likewise, experienced mathematics teachers were often able to pre-empt pitfalls and hence preferred examples that can lead students away from forming misconceptions (Bills & Bills, 2005).

While acknowledging the fact that the teachers' knowledge influences their use of examples, this paper focuses on the following three questions.

(1) How do secondary mathematics teachers choose examples for introducing new mathematical ideas?
(2) How do secondary mathematics teachers select homework task(s)?
(3) What are the characteristics of a "good" mathematical example in secondary teachers' perceptions?

2 Mathematical Examples

The use of examples by teachers in the teaching of mathematics are so entrenched in Mathematics education, that Bills and Watson (2008, p. 77) claimed that "any theory of learning which does not deal with how learners and teachers act with, and on, examples is likely to be incomplete as far as mathematics is concerned". The significance of examples is also summarized in what Watson and Mason (2002, p. 39) wrote, "learning mathematics can be seen as a process of generalizing from specific examples". Examples are therefore paramount in mathematical teaching and learning.

The definition of examples used by researchers although not completely identical, generally refers to an example as an illustration of a larger class. This broad definition can include geometrical figures, demonstrations of solving problems, tasks, worked examples, as long as the mathematical object is offered or perceived as an example of something. In this study, a task can be an exercise, problem or assessment, assigned to students for completion during or beyond curriculum time. The same task may differ in operation and learning outcomes, depending on the intentions of the author, the aims and knowledge of the teacher, the goals, knowledge, and experiences of the students, and on the learning environment. As stressed by Cai (1997, p. 9), "the use of a variety of mathematical tasks can capture a range of students' mathematical performance". The role of teachers therefore lies in planning and setting appropriate tasks.

Example selection is however not merely choosing or implementing good examples, it entails leveraging on coherent example sets that are purposefully sequenced and varied in order to gradually build students' understanding to attain instructional goals. The variation theory (see Marton & Booth, 1997) posits that learning means to experience an object of learning in a novel or different way. In order to experience, one needs to be able to discern the critical aspects of the object of learning, which can be brought to the learner's awareness through planned variation. Adapting the notion of variation into exemplification, Watson and Mason (2006) claimed that "the starting point of making sense of any data is the discernment of variations within it" (p. 92). They proposed to systematically change certain aspects of a task while keeping others invariant, to help learners better perceive the mathematical structure. They warned that if too many attributes are changing at the same time, it will be impossible for the learner to abstract the essential features, reducing a series of meaningful examples into mere disconnected and unstructured isolated examples.

In addition, Skemp (1971) advised educators to reduce the 'noise' in examples or the peripheral attributes of the concept during concept formation so as to draw learners' attention to the key characteristics of the concept. When strengthening the conceptual understanding, Skemp (1971) proposed educators to increase the noise in examples so as to heighten learners' ability in pinpointing the key properties which are now made more obscure. Non-examples have also been proposed by researchers in the teaching of mathematical concepts to prevent over-generalization of the concepts (Petty & Jansson, 1987).

Empirical findings from work with teachers have also revealed principles that guide teachers in making their example choices. One common approach was the use of *simple first examples* (Bills & Bills, 2005; Zodik & Zaslavsky, 2008) in the development of a mathematical argument or in the understanding of a procedure, which includes keeping the numbers small and ordering examples in increasing complexity. Nevertheless, the advice of "keeping things simple" must not be taken at face value as the subsequent message to provide systematic variation and more complicated examples so that learners would not under-generalize is equally important. On the contrary, randomness is sometimes deemed by some teachers as a valid option in "conveying the idea of generality" (Zodik & Zaslavsky, 2008, p. 175), for example, through the use of randomly selected coefficients. However, studies have also forewarned that randomly generated examples may be ineffective and might even be misleading.

To scaffold students' learning, teachers have also proposed using examples that build on students' prior knowledge and are ordered in increasing complexity (Bills & Bills, 2005) and keeping unnecessary work to a minimum (Zodik & Zaslavsky, 2008). Sometimes, teachers tend to craft and use examples that allow them to attend to common errors and misconceptions to alert their students (Zodik & Zaslavsky, 2008) or to include uncommon cases to increase students' exposure.

3 Teacher Knowledge

Before focusing our attention on the relationship between teachers' knowledge and their exemplification behavior, it is worthwhile to examine teachers' knowledge. This study uses the definition by Ball, Thames and Phelps (2008, p. 399), whereby teacher knowledge is described as the mathematical knowledge "entailed by teaching". In other words, teacher knowledge is required to perform the task of teaching mathematics to students. In particular, content knowledge and pedagogical content knowledge (PCK) have been surfaced to directly influence teachers' exemplification abilities.

Content knowledge relates to the amount and organization of subject matter knowledge in one's mind. It comprises of substantive knowledge or knowledge of facts, concepts, principles, definitions, theorems, rules and structures (Shulman, 1986). Beyond that, content knowledge also consists of syntactic knowledge which involves understanding the rules of proof and the methods of enquiry into the domain. Therefore, "the teacher need not only understand that something is so; the teacher must further understand why it is so" (p. 9). Although it is not within the scope of this paper to investigate the impact of teachers' content knowledge on their example use, it is important to keep in mind the connection between the two.

Shulman (1986, p. 8) described PCK as the "blending of content and pedagogy into an understanding of how particular topics, problems, or issues are organized, represented, and adapted to the diverse interests and abilities of learners, and presented for instruction". It is the knowledge that allows one to frame and communicate mathematics in a manner to make learning easier for the learner. Since the publication of Shulman's seminal paper, PCK has generated much interest in the research community and researchers have subsequently refined, developed and tested fine-grained conceptualizations of PCK. Leikin and Zazkis (2010, p. 454) extended Shulman's PCK to include the "awareness of the cognitive, social and affective characteristics of a mathematics classroom". In the Teacher Education and Development Study in Mathematics (TEDS-M), the term mathematics pedagogical content knowledge (MPCK) which relates to the PCK for mathematics teaching (Tatto et al., 2012) was used. Three sub-domains were identified: curricular knowledge, planning for teaching and learning, and enacting teaching and learning.

Building on Shulman's work, Ball et al. (2008) developed a framework that divides teachers' mathematical knowledge for teaching into subject matter knowledge and PCK, and this is used to frame the current study. Under the label of subject matter knowledge, Ball and her colleagues identified common content knowledge (CCK), specialized content knowledge (SCK) and horizon content knowledge (HCK). CCK relates to mathematical knowledge and skills used in a wide variety of settings, not unique to teaching, whereas SCK refers to those unique to teaching. HCK is defined as "an awareness of the large mathematical landscape in which the present experience and instruction is situated" (Ball & Bass, 2009).

Ball et al. (2008) sub-divided PCK into knowledge of content and students (KCS), knowledge of content and teaching (KCT), and knowledge of content and curriculum (KCC). KCS includes an awareness of topics that students will find easy or difficult and their common conceptions and misconceptions. KCT comprises of knowledge on the sequencing of examples and the use of appropriate representations. Finally, KCC encompasses knowledge of educational goals, assessments, and the sequencing of topics across grade levels.

4 Teacher Knowledge and Mathematical Examples

A closer scan of the literature on mathematical examples highlights the close connection between teachers' examples and their knowledge. In particular, content knowledge and PCK were pinpointed to directly influence teachers' exemplification abilities.

Rowland et al. (2005) observed how content knowledge and PCK guided the instructional decisions and actions of prospective elementary teachers in their UK classrooms during school placement. Through the video analysis of the teacher trainees' lessons, Rowland and his fellow researchers developed the Knowledge Quartet framework as a tool for thinking how subject knowledge plays out in the classroom. Of the four units of the Knowledge Quartet framework, transformation or knowledge-in-action was tightly linked to teachers' example choice, which was significantly prevalent in the trainees' lessons. Variables, sequencing, representations, and learning objectives were also identified to be related to teachers' awareness in exemplification (Rowland, 2008).

Noticing the lack of research between teachers' PCK and their exemplification practices, Chick and her colleagues studied the instructional practices of Australian elementary teachers and were successful in locating moments where aspects of PCK were enacted through the teachers' examples. Chick (2007) noted that most of the examples that the teachers used were planned and selected based on the examples' structures and qualities. The selection process was much guided by the teachers' PCK, especially on what affordances they perceived the examples could offer. Even when teachers have to come up with an example on the spot, their ability to do so is greatly influenced by their PCK (Chick & Pierce, 2008). A distinction was made too, between selecting examples and using them as "even well-chosen examples are not necessarily easy to implement effectively in the classroom" (Chick & Pierce, 2008, p. 321).

Similarly, Zodik and Zaslavsky (2008) who carried out an in-depth study that specifically addressed mathematics teachers' use of examples surfaced many novel findings, establishing exemplification as a rich research domain where much remains unexplored. For instance, they reported that the ratio of spontaneous examples to planned examples were almost equal, hence emphasizing the importance of teachers' knowledge in crafting instructional examples during lesson delivery. In both cases, whether the examples were planned or otherwise, the amount of deliberation that went into creating or selecting and implementing each example underscored the complexity in exemplification. From their interviews with and lesson observations of five experienced secondary teachers, they concluded that content knowledge, PCK, and knowledge of students' learning, a sub-category of PCK, were fundamental in shaping teachers' examples.

In the area of mathematical tasks, Sullivan, Clarke, Clarke and O'Shea (2009) examined how the enactment of the same task varied in the classrooms of three teachers and the findings asserted the notion that execution of tasks is shaped by the instructional goals and knowledge of the teachers. Thus, the ability to convert tasks into effective lessons relies on teachers' PCK and literature has again revealed teachers' knowledge as the crux of their exemplification abilities.

5 Methodology

This study examined the exemplification practices of secondary mathematics teachers in Singapore. As the research aim is to examine the interconnectedness between teachers' knowledge and their examples, it makes more sense to survey experienced teachers for two reasons. First, teachers who have taught mathematics for a substantial number of years would have rehearsed and refined their exemplification skills countless times in their classrooms. They would therefore have a wider repertoire of inter-related mathematical examples that can easily be brought to the fore, backed by pedagogical causes. In contrast, beginning teachers are likely to have a heavier reliance on stipulated curriculum materials while they are still grappling with issues both within and beyond the classrooms. Hence, a purposeful sample of experienced mathematics teachers was imperative. As the main study was set up in the form of multi-case studies, a questionnaire was constructed to locate the key participants.

Participants were to be chosen from teachers who had taught mathematics for at least five consecutive years and had some experience in teaching at the upper secondary level. Approval was then sought from the Ministry of Education (Singapore) and thereafter from the schools' principals before the questionnaire was pilot-tested with 16 teachers from two schools who met the criteria, and subsequently refined. An information sheet and a consent form were attached to each questionnaire to inform participants of the objectives of the teacher questionnaire and to seek their consent. There were also written instructions on the questionnaire that appealed to teachers to refrain from discussing the contents of their questionnaire with anyone to ensure its validity. The open-ended questions in the questionnaire were general about the teaching of mathematics and were not tied to any specific content domain in mathematics.

A total of 128 teachers from 24 secondary schools responded to the questionnaire, with a good spread of teachers from five to seven schools in each of the four zones of Singapore. Seven of the returns were deemed invalid as three of the teachers had only lower secondary (grade 7 and grade 8) teaching experience and four had taught for less than five years. The remaining 121 teachers had a mean of 12 years of teaching experience and 89 of them had experience in teaching Additional Mathematics, an advanced level of mathematics that is offered to more mathematically-inclined students in upper secondary. Of these 121 teachers, 44 teachers taught one other subject and the rest, nearly two-thirds of the teachers taught mathematics only. All respondents had a first degree and a teaching qualification. 25 of the teachers had a master's degree of which 19 were masters in mathematics or mathematics education. The gender composition was almost 50:50 (57 females). 119 indicated their age group and the age distribution is shown in Table 1.

Table 1

Age group of 119 teacher respondents

Age	Under 30	30 – 39	40 – 49	50 – 59	60+
Number of teachers	7	58	32	17	5

The purpose of this questionnaire was to explore the teachers' opinions on mathematical examples, their mathematical knowledge of teaching, and their mathematical beliefs. For this paper, the focus is on the three open-ended questions that surveyed the teachers' exemplification practices. The first question read "list down two factors you consider when selecting examples to introduce a new concept/procedure/rule/principle". Research has shown that teachers like to begin with a simple or familiar first example and order examples in increasing degree of difficulty (Rowland et al., 2005). Teachers also reported to be conscious of the importance to reduce the 'noise' in examples so as to focus learners' attention on the critical aspects (Skemp, 1971). Hence, the objective of this question was to elicit teachers' decisions in selecting their first examples in order to focus on those teachers who can better justify their choice of mathematical examples.

The second question asked teachers to list down two factors they considered when selecting homework tasks. Homework is a common feature in the teaching of mathematics in Singapore schools. Hiebert et al. (1996) proposed that teachers look for three essential characteristics in tasks. Tasks that can offer situations that students will perceive as problematic and second, tasks that provide platforms for students to think about important mathematics. Third, tasks should also connect to some part of the students' knowledge so that they are attainable by students. The researchers added that "the teacher will need to take an active role in selecting and presenting tasks" (p. 16). Hence, it is worthwhile to investigate how teachers decide on homework tasks.

Finally, teachers were asked to write down three characteristics of what they think a "good" example would have, since teachers' examples are picked from a pool of countless possibilities, where some are simply more appropriate than others. Zaslavsky and Lavie (2005, p. 2) defined a good example as one "that conveys to the target audience the essence of what it is meant to exemplify or explain". First, good examples must be transparent, one that aims to accentuate the critical features while making the irrelevant attributes as subtle as possible, so as to allow learners' attention to be drawn to the underlying structure. Second, good examples should foster generalization, such that the learners can see "the generality embodied in the example" (Mason & Pimm, 1984, p. 286). The third example trait is such that these examples should aid in explaining and resolving mathematical subtleties. Thus, the third question was to elicit what teachers in Singapore believed that a good example would entail.

6 Results and Discussion

The data collected for this study focused on teachers' responses to the three questions. Teachers' responses for each question were categorized and 13 category codes were created to facilitate the analysis and discussion both within and between the questions. Table 2 presents the percentage category frequencies for each question, ordered in decreasing frequencies for question one.

6.1 *How do secondary mathematics teachers choose examples for introducing new mathematical ideas?*

A total of 235 teachers' considerations, when they teach new mathematical ideas, were gathered in which the first three categories emerged more often. From Table 2, *Student Abilities* (SA) was reported as the major concern teachers have when introducing new content (60 counts). SA consisted of responses on students' abilities, prior knowledge, and the need to scaffold students' learning. The comments included "must suit students' ability", "pitched to students' current level of understanding", or simply "ability appropriate". When making example choices, these teachers pay special attention to students' "pre-requisite knowledge" so as to "link to prior knowledge" or "build on concepts that they have already learnt". Some teachers, like the mentors in Bills and Bills' (2005) study, also advocated instructional scaffolding

Table 2
Categories of 121 teachers' exemplification considerations

Category Code	Category Description	Teach Mathematics Idea (%)	Select Homework (%)	Good Example (%)
SA	Students' Abilities	25.5	17.4	13.1
DL	Difficulty Level	21.3	23.0	16.1
FC	Familiar Context	18.3	–	8.36
LO	Learning Objectives	8.09	8.12	5.97
EC	Exemplify Content	8.09	–	10.7
VE	Variety of Examples	6.81	19.2	10.1
CE	Clarity of Examples	5.11	–	15.8
TI	Thinking and Interesting	3.82	–	9.25
CM	Common Misconceptions	2.13	0.855	4.18
CH	Classwork and Homework	0.851	5.98	–
NE	Number of Examples	–	9.83	–
RL	Reinforce Learning	–	8.94	4.78
AU	Assess Understanding	–	6.41	1.49

via examples like "easy ones first, then progressively more challenging ones" or "try to start with numerical examples then go into algebra to generalize the learning". This deliberation in the sequencing examples was also observed in the elementary teacher trainees in Rowland's (2008) study.

The second most common category was *Difficulty Level* (DL) which pertains to whether the examples should be easy or hard (50 counts). Note that for the second and third research questions, there were teachers who preferred to include challenging examples to stretch their students, but, this was not the case when teachers teach new mathematical ideas. Slightly over three-fifths proposed to use an example that is "basic", "straight-forward and easy to understand" so that it "does not confuse students" and this resembles the key theme in another study which was to keep things simple (Bills & Bills, 2005). The rest echoed that they would take into account the "level of difficulty" of the first examples. A related category with 43 counts was to use *Familiar Context* (FC) that students can easily relate to by linking to the "personal experience of students" or "relating to their daily lives if possible". Examples that were "authentic" or offer "real-world situations" were also raised.

Of the participating teachers, 19 touched on the "relevance of the example" or if it is "linked to concept", whether the examples could "address the instructional objectives" and prepare students for examinations (LO). Of which, three explicitly stated that assessment guided their example decisions, such that they will seek examples with "relevance to O Level exam questions". This factor was also cited by Rowland (2008) in his study. Teachers were equally mindful when selecting the first few examples that could exemplify a new content (EC) clearly, with "not too many variation" but "should convey the essentials of concepts involved". Some of the comments included "examples must clearly demonstrate only the concept/procedure/rule/principle", present the "general principle, not special case", and are able to "highlight the key points". This is somewhat reflective of teachers' intent to reduce the noise (Skemp, 1971) of an example or to use transparent examples (Zaslavsky & Lavie, 2005).

There were 16 comments on using different examples, *Variety of Examples* (VE), when presenting a new mathematical idea or to "choose a variety of standard versus non-standard examples" or those that "show the application of the new concept". Again, this is coincident with what literature has suggested to provide students with varied experiences in order for them to attend to what has changed and what has remained the same (Rowland, 2008).

There were 12 teachers who wrote about the *Clarity of Examples* (CE) and that examples should be clear, "should not be overly tedious to solve" nor "be clouded by other concepts or difficult algebraic manipulation". Examples should involve "small numbers, positive integers if possible" or "numbers that are not intimidating". This partially reflected the approach by teachers in another research to draw attention to relevant features (Zodik & Zaslavsky, 2008).

Arousing interest and stimulating thought processes, *Thinking and Interesting* (TI) were also raised (9 counts). Fewer (5 counts) attended to the need to address *Common Misconceptions* (CM) where one added to "highlight non-examples to check on understanding". Only two teachers mentioned that they will select examples that "can help them (students) to solve questions given for homework later" (*Classwork and Homework*-CH). Since the teaching of a new mathematical idea was the focus of this question, it was logical that the following categories: *Number of Examples* (NE), *Reinforce Learning* (RL), and *Assess Understanding* (AU), were not part of the teachers' considerations.

In summary, the top three categories, SA, DL, and FC, which made up more than 60% of the teachers' considerations, were closely related. The key theme that threaded through these three categories was to present students with an easy (DL) or familiar (FC) example when introducing a new mathematical idea, so that the example is within students' reach (SA). In a way, this key theme encompassed one of the guiding principles teachers in Zodik and Zaslavsky's (2008) study exhibited, which was to begin with a simple or familiar case, or that in another study, "keeping things simple" (Bills & Bills, 2005).

6.2 *How do secondary mathematics teachers select homework task(s)?*

Homework tasks included exercises, problems and other assessment items given to students. There were 234 written factors that surfaced as responses for this question form the teachers. The top three categories, DL, VE, and SA were frequently cited by many of the respondents. Similar to teachers' choice of the first few examples, when they plan homework, DL (54 counts) and SA (41 counts) were important too. What differs in DL when planning homework tasks as compared to teaching new mathematical ideas, was that teachers preferred a "mixture of easy and difficult" examples and more were more prone to choose complex over simple homework tasks or to include "1 or 2 higher-order questions to challenge students". "Tasks should be reasonable within ability of students" so that "students can manage the homework". Hiebert et al. (1996, p. 16) considered SA as vital too as teachers should select and scaffold tasks that "students can see the relevance of the ideas and skills they already possess".

A strategic approach by many (45 counts) was to provide students with varied examples (VE), as a limited range of examples might lead to an incomplete or erroneous understanding. "Direct application of concepts, challenging questions, and integrated mathematics and real-life situations" should be tasked for a "comprehensive coverage of exercise". In addition, teachers tend to "expose students to different ways of questioning" or to "involve the same concept in different contexts". VE was thus viewed more crucial when choosing homework tasks than first examples.

The next three codes, NE (23 counts), RL (21 counts), and LO (19 counts) had comparable ratings. Some teachers carefully considered the "time taken to complete homework questions" by reminding themselves to give "manageable number of questions" (NE). Hence although some studies which have looked into Singapore classrooms have observed that teachers tend to assign students with a large number of practice tasks (Kaur, 2010), this study showed that teachers do take into account the amount of time students spent on their homework. However, this category was absent in the teachers' exemplification deliberations when they introduced new concepts or when they identified good examples. Some teachers were concerned whether homework could "reinforce classroom teaching" (RL) and help students "to acquire proficiency". The "purpose of the homework task" (LO) to cover the school's "scheme of work" or to "prepare students for examinations" was also raised. This corresponded with part of the findings in another study where it was reported that teachers in Singapore sometimes select practice questions which were likely to be tested (Fang, Ho, Lioe, Wong, & Tiong, 2009).

15 teachers suggested that the role of the homework is "to assess students' understanding" (AU) and that "tasks should give feedback on students' learning" or "aids students in consolidating their learning". Slightly fewer (14 counts) shared that their homework selection was based on the classwork and that for the homework they "will give questions similar to the work done in class" or "similar to teaching examples". Lastly, only two stated that they would include "questions that can surface common mistakes or misconceptions" (CM).

It was noticeable that FC, EC, CE, and TI were not factored in when teachers set homework tasks. Since homework served mainly for students to develop their skills and to apply the concepts learnt, teachers reported that they tended to expose students to different types of problems. Thus, categories like EC and CE which were more relevant to the forming of mathematical understanding were not part of teachers' guiding principles in setting homework tasks. A sharp contrast was evident too in the use of familiar context (FC), which although was the third top guiding principle when introducing new mathematical ideas, was deemed unnecessary by experienced teachers when choosing homework tasks. What was more conspicuous were the absence of references to higher order thinking and interesting contexts in the homework tasks, as these are deemed fundamental in the Singapore mathematics curriculum framework (Ministry of Education, 2012).

To summarize, when choosing homework tasks, teachers pay most attention to the difficulty level of tasks (DL), offer varied tasks to students (VE) and ensure that the tasks suits their students' ability level (SA). An assortment of examples with varying difficulty level is dished out for homework where the routine ones are meant for practice (Rowland, Thwaites, & Huckstep, 2003; Watson & Mason, 2004) and the demanding ones are to extend learning (Hiebert et al., 1996). Similarly, besides varying the difficulty level, ensuring a diversified array of examples is also essential to increase student' example space and to provide them with a fuller experience. This study also surfaced what the teachers valued in homework tasks, which is not just to reinforce and facilitate students' learning, but also to prepare students for assessment.

6.3 *What are the characteristics of a "good" mathematical example in secondary teachers' perceptions?*

The respondents gave 335 written descriptions of their concept of good examples. Likewise, when teachers look for critical attributes in examples, DL (54 counts) and SA (44 counts) were pivotal. Interestingly, over 75% were more likely to pick an "easy to understand", "basic" or "direct" example over one that "can stretch their thinking". For teachers who were more concerned with their students' abilities (SA), they felt that a good example should be "pitched at the right level for the class" and be able to "link with prior knowledge". Slightly more than half in this category identified with the scaffolding function of examples where a good example set should progress "from direct to more complicated" or in the form "concrete-pictorial-abstract". Unlike the previous two questions, there were five teachers who favored the use of "illustrations and diagrams" to "assist in the conceptualization", which Rowland et al. (2005) found to be tied to teachers' exemplification practices.

A substantial number of teachers (53 counts) described good examples as "clear" (CE), "concise" with "no ambiguity" and "well-crafted", where they "test students on the concept but not on the English". "Ease in calculation" and having "no complicated equations" reflected the keep unnecessary work to a minimum strategy, discussed earlier in Zodik and Zaslavsky (2008). This category, although highly regarded by teachers in choosing good examples was nonetheless missing from teachers' list of homework task characteristics and also not ranked high in their selection of first examples.

Teachers (36 counts) also characterized those that are "representative", able to "highlight the salient points" (EC) and enable one "to generalize ideas or rules" as good examples. Hence, good examples are transparent and promote generalization (Zaslavsky & Lavie, 2005). Others (34 counts) deemed it a necessity for examples to be "varied" (VE) to provide "sufficient coverage", to "link concepts together", and to allow the "application of concepts across topics".

Another desirable attribute of an example is if it is "able to provoke thinking", "involve inference", "arouse students' interest" or "able to spark discussion" (TI). Of this type, 31 counts were identified and we can draw a parallel between TI and what Hiebert et al. (1996, p. 18) meant by tasks that problematized the subject, so that they will "pique the interests of students and engage them in mathematics". Following next, is teachers' preference (28 counts) for examples "related to everyday experiences of students" (FC) or "has real-life application".

Twenty teachers indicated that a good example "delivers the lesson objectives" (LO) and some felt that it should be "similar to the examination syllabus type of questions". Lesser (16 counts) highlighted examples that "reinforce concepts or skills taught in class" (RL) to "allow students to understand the concept better". 14 felt that good examples offer "opportunities to sieve out misconceptions in students" (CM) so as to attend to students' errors (Zodik & Zaslavsky, 2008). There were only five comments on choosing examples that can "provide good feedback about students' understanding" (AU). Finally, it is self-explanatory that categories like CH and NE are absent from teachers' conception of good examples.

The responses about DL are reflective of teachers' considerations at different stages of instruction, easy at first, and thereafter harder ones to deepen students' understanding. In the discussion for a good example, clarity of the example was highly regarded so as to allow "one to see the general through the particular" (Mason & Pimm, 1984). This is also consistent with researchers' recommendation of "transparent" examples to convey the essence of what the example meant to exemplify or explain (Zaslavsky & Lavie, 2005). Overall, DL and SA remained pivotal when teachers selected examples (across the three RQs).

6.4 *Teacher knowledge*

The three questions discussed in this paper were not based on any specific mathematical content. However, another section of the questionnaire examined teachers' mathematical knowledge. The data here suggested that there were obvious connections between teachers' PCK and their use of examples, namely KCS, KCT and KCC.

When teachers present new content, KCS is exhibited in how they considered students' prior knowledge (SA) and the difficulty level (DL) of the topic. As such the teachers try to choose ability-appropriate examples that students can relate to (FC) and find interesting (TI) to make learning more manageable and meaningful for the students. Furthermore, knowledge of students' conceptions and misconceptions (CM) means that teachers prefer examples "that should not be clouded by other concepts or difficult algebra manipulation" (CE) so as not to confuse their students (Ball et al., 2008). Each of the above-mentioned categories requires teachers' knowledge of how students learn the mathematical content or KCS in short.

Teachers' example choice is influenced by their KCT too. They select examples that are able to exemplify the mathematical idea (EC) and also provide students with sufficient contact with the mathematical content through varied examples (VE). Teachers' KCT guide them in the sequencing of homework tasks in "ascending difficulty" (SA) in order to scaffold students' learning. In addition, teachers tend to pick those tasks that are able to reinforce what has been taught (RL) or by relating homework tasks to what have been covered in class (CM), in order to help students retain knowledge and gain fluency in their mathematical competency (Rowland, 2008). Challenging tasks (DL) are also utilized to bring students deeper into the topic.

Finally, teachers' knowledge of the curriculum (KCC) sensitize them to those examples that are able to address and deliver learning objectives stipulated in the mathematics syllabus, as well as prepare students for assessment (LO) by making available to them examples that are similar to those tested in examinations. At the same time, teachers leverage on examples that "provide good feedback about students' understanding" (AU) in order to improve students' learning.

7 Conclusion

Mathematics is one of the key content areas for the 21st century (Partnership for 21st Century Learning, 2016) and mathematics teachers have the important responsibility of educating the next generation of learners in schools. In addition, one of the claims by MOE underpinning the 21st century competencies is that: "Knowledge and skills must be underpinned by values" (MOE, 2010). The teachers in this study did not bring up the idea of "values" as one of the criteria useful for selecting examples for instruction. However, we should be careful in concluding that our teachers are not considering values in their teaching. This study provides baseline data about what mathematics teachers value when they select examples for instruction and this information can be useful for organizing PD courses for teachers.

Teachers will continue to use examples in teaching their students, for whom examples may be a primary means for learning mathematical concepts. The use of certain examples for teaching a particular topic may not be universal, which implies that the survey of the teachers from Singapore who participated in this study may be very context-specific. Moreover, "the choice of an example for teaching is often a trade-off between one limitation and another" (Zaslavsky, 2014, p. 29).

It is important to be aware of the limitations in using questionnaire findings to study teachers' pedagogical practices since what is written may not be translated in actual lessons. Also, we have to be mindful of the fact that what is written by the teachers may not necessarily translate into actual classroom practice. Nevertheless, this study brings us some insights into the exemplification perceptions of experienced mathematics teachers in Singapore. Teachers are most concerned over students' abilities and the difficulty level of examples when choosing examples. However, when selecting examples for different purposes, the considerations differ to some extent. For instance, when introducing new content, teachers favored examples that connect with students' experiences whereas for homework, they are more concerned with providing students with varied exposure. Example selection by Singapore teachers seem to be guided by instructional considerations and point to a connection between teacher knowledge and instructional examples.

Finally, this research reveals the potential direction for further research into the reasons teachers considered as critical factors in their choice of examples and points to a connection between teacher knowledge and beliefs about what constitutes effective teaching and learning of mathematics through the use of mathematical examples.

References

Ball, D. L, & Bass, H. (2009). With an eye on the mathematical horizon: Knowing mathematics for teaching to learners' mathematical futures. *Paper presented at the 43rd Jahrestagung fur Didaktik der Mathematik*, Oldenburg, Germany.

Ball, D. L., Thames, M. H., & Phelps, G. (2008). Content knowledge for teaching: What makes it special? *Journal of Teacher Education*, *59*(5), 389–407.

Bills, C., & Bills, L. (2005). Experienced and novice teachers' choice of examples. In P. Clarkson, A. Downton, D. Gronn, M. Horne, A. McDonough, R. Pierce, & A. Roche (Eds.), *Proceedings of the 28th Annual Conference of the Mathematics Education Research Group of Australasia* (Vol. 1, pp. 146–153). Melbourne, Australia: MERGA.

Bills, L., & Watson, A. (2008). Editorial introduction. *Educational Studies in Mathematics*, *69*(2), 77–79.

Cai, J. (1997). Beyond computation and correctness: Contributions of open-ended tasks in examining U.S. and Chinese students' mathematical performance. *Educational Measurement: Issues and Practice*, *16*(1), 5–11.

Chick, H. L. (2007). Teaching and learning by example. In J. Watson & K. Beswick (Eds.), *Mathematics: Essential research, essential practice. Proceedings of the 30th Annual Conference of the Mathematics Education Research Group of Australasia* (pp. 3–21). Sydney, Australia: MERGA.

Chick, H. L. (2010). Aspects of teachers' knowledge for helping students learn about ratio. In L. Sparrow, B. Kissane, & C. Hurst (Eds.), *Shaping the future of mathematics education. Proceedings of the 33rd Annual Conference of the Mathematics Education Research Group of Australasia* (pp. 145–152). Fremantle, Australia: MERGA.

Chick, H. L., & Pierce, R. (2008). Issues associated with using examples in teaching statistics. In O. Figueras, J. L. Cortina, S. Alatorre, T. Rojano, & A. Sepúlveda (Eds.), *Proceedings of the Joint Meeting of PME 32 and PME-NA XXX* (Vol. 2, pp. 49–55). Mexico: Cinvestav–UMSNH.

Fang, Y., Ho, K. F., Lioe, L. T., Wong, K. Y., & Tiong, Y. S. J. (2009). *Developing the repertoire of heuristics for mathematical problem solving: Technical Report for Project CRP1/04 TSK/JH.* Singapore: Centre for Research in Pedagogy and Practice, National Institute of Education, Nanyang Technological University.

Hiebert, J., Carpenter, T. P., Fennema, E., Fuson, K., Human, P., Murray, H., Olivier, A., & Wearne, D. (1996). Problem solving as a basis for reform in curriculum and instruction: The case of mathematics. *Educational Researcher*, *25*(4), 12–21.

Huntley, R. (2013). Pre-service primary teachers' choice of mathematical examples: Formative analysis of lesson plan data. In V. Steinle, L. Ball, & C. Bardini (Eds.), *Mathematics education: Yesterday, today and tomorrow. Proceedings of the 36th annual conference of the Mathematics Education Research Group of Australasia* (pp. 394–401). Melbourne, Australia: MERGA.

Kaur, B. (2010). A study of mathematical tasks from three classrooms in Singapore. In Y. Shimuzu, B. Kaur, R. Huang, & D. Clarke (Eds.), *Mathematics tasks in classrooms around the world* (pp. 15–33). Rotterdam: Sense Publishers.

Leikin, R., & Zazkis, R. (2010). On the content-dependence of prospective teachers' knowledge: A case of exemplifying definitions. *International Journal of Mathematics Education in Science and Technology, 41*(4), 451–466.

Marton, F., & Booth, S. (1997). *Learning and awareness.* Mahwah, NJ: Lawrence Erlbaum.

Mason, J., & Pimm, D. (1984). Generic examples: Seeing the general in the particular. *Educational Studies in Mathematics, 15*(3), 277–289.

Ministry of Education (2010). *MOE to enhance learning of 21st century competencies and strengthen art, music and physical education.* Retrieved 31 December, 2015 from www.moe.gov.sg

Ministry of Education (2012). *O & N(A) Level mathematics teaching and learning syllabus*. Singapore: Curriculum Planning and Development Division.

Partnership for 21st Century Learning. (2016). *Framework for 21st Century Learning.* Retrieved 1 January, 2016 from http://www.p21.org/storage/documents/docs/P21_framework_0116.pdf

Petty, O. S., & Jansson, L. C. (1987). Sequencing examples and nonexamples to facilitate concept attainment. *Journal for Research in Mathematics Education, 18*(2), 112–125.

Rowland, T. (2008). The purpose, design and use of examples in the teaching of elementary mathematics. *Educational Studies in Mathematics, 69*(2), 149–163.

Rowland, T., Thwaites, A., & Huckstep, P. (2003). Elementary teachers' mathematics content knowledge and choice of examples. *Proceedings of the third conference of the European Society for Research in Mathematics Education*, Italy.

Rowland, T., Huckstep, P., & Thwaites, A. (2005). Elementary teachers' mathematics subject knowledge: The knowledge quartet and the case of Naomi. *Journal of Mathematics Teacher Education, 8*(3), 255–281.

Shulman, L. S. (1987). Knowledge and teaching: foundations of the new reform. *Harvard Educational Review, 57*(1), 1–22.

Skemp, R. R. (1971). *The psychology of learning mathematics.* Harmondsworth, UK: Penguin.

Sullivan, P., Clarke, D., Clarke, B., & O'Shea, H. (2009). Exploring the relationship between tasks, teacher actions, and student learning. In M. Tzekaki, M. Kaldrimidou, & H. Sakonidis (Eds.), *In search of theories in mathematics education.*

Proceedings of the 33rd Conference of the International Group of Psychology of Mathematics Education (Vol. 5, pp. 185–192). Thessaloniki, Greece: PME.

Tatto, M. T., Schwille, J., Senk, S. L., Ingvarson, L., Rowley, G., Peck, R., Bankov, K., Rodriquez, M., & Reckase, M. (2012). *Policy, practice, and readiness to teach primary and secondary mathematics in 17 countries. Findings from the IEA Teacher Education and Development Study in Mathematics (TEDS-M)*. Amsterdam: IEA.

Watson, A., & Mason, J. (2002). Extending example spaces as a learning/teaching strategy in mathematics. In A. Cockburn & E. Nardi (Eds.), *Proceedings of the 26th Conference of the International Group for the Psychology of Mathematics Education* (Vol. 4, pp. 378–385). Norwich: PME.

Watson, A., & Mason, J. (2004). The exercise as mathematical object: Dimensions of possible variation in practice. In O. McNamara (Ed.), *Proceedings of the British Society for Research into Learning Mathematics* (Vol. 24, No. 2, pp. 107–112). Leeds, UK: BSRLM.

Watson, A., & Mason, J. (2006). Seeing an exercise as a single mathematical object: Using variation to structure sense-making. *Mathematical Thinking and Learning, 8*(2), 91–111.

Zaslavsky, O. (2014). Thinking with and through examples. In C. Nicol, P. Liljedahl, S. Oesterle, & D. Allan (Eds.), *Proceedings of the Joint Meeting of PME 38 and PME-NA 36* (Vol. 1, pp. 21-34). Vancouver, Canada: PME.

Zaslavsky, O., & Lavie, O. (2005). *Teachers' use of instructional examples*. Paper presented at the 15th ICMI Study Conference, Águas de Lindóia, Brazil.

Zodik, I., & Zaslavsky, O. (2008). Characteristics of teachers' choice of examples in and for the mathematics classroom. *Educational Studies in Mathematics, 69*(2), 165–182.

Chapter 12

On the Efficacy of Flipped Classroom: Motivation and Cognitive Load

Weng Kin HO   Puay San CHAN

In the turn of the 21st century, there has been growing evidence that suggest the possibility of replacing traditional transmissive lecture or direct teacher instruction with active in-class activities and pre- or post-class work via the so-called 'flipped classroom pedagogy'. By a modification of the theoretical framework proposed by Jacob Lowell Bishop and Matthew Verleger in 2013, we describe an authentic implementation of flipped classroom for teaching Mathematics to a selected group of junior college Year 2 students (12th grade) in Singapore. The effectiveness of our implementation is then analysed through five of the six testable propositions put forward by Lakmal Abeysekera and Phillip Dawson in 2015.

1 Introduction

In the opening years of the 21st century, teaching approaches that advertise the success of the use of flipped classroom have attracted an increasing amount of attention. Common to most of these implementations is the movement of the information-transmission component of traditional in-person lecture or teaching time (referred to as traditional classroom subsequently) *out of class time*. Prior to coming for class, students are typically given resources that scope what would have been covered in the lecture, alongside with exercises or activities intended for independent completion as *pre-class activities*. Classroom time is then spent on *in-class activities* including quizzes to measure students' basic level of mastery of the content knowledge acquired prior the in-class sessions, as well as group discussions centred about problem solving and applications of concepts acquired earlier. Lage, Platt and Trelia (2000) pioneered the idea of 'inverted classroom'. But the first scholarly use of the word 'flipped' classroom appeared in Strayer's (2007) doctoral dissertation on the topic. Since then, there has never been a universally agreed definition. Lage et al.'s definition that "inverting the classroom means that events that have traditionally taken place inside the classroom now takes place outside the classroom and vice-versa" (Lage et al., 2000, p. 32) has become obsolete as current research in flipped classroom is moving away from a mere label of re-ordering classroom and at-home activities.

1.1 *Definition of flipped classroom adopted in this chapter*

As 'flipping' entails the inversion of expectations in traditional lecture (Berrett, 2012), we exploited the use of computer technology and the Internet (e.g., video-recorded lecture available online) to move the information-transmission component of a traditional lesson out of class time and replaced by a spectrum of scaffolding activities designed to motivate independent learning. Thus, our working definition of flipped classroom can be seen as an educational method comprising two sub-systems: interactive learning system inside the classroom, and a direct computer-based instruction outside the classroom. We emphasise that video technology is harnessed here because pedagogical theories of grounded image (Ho, Leong & Ho, 2015) have informed us that students are better able to discuss, with specific reference to particular juncture of the video-recorded lectures, the content knowledge transmitted with their peers and teachers. Consequently, in this chapter, we restrict our definition of 'flipped classroom' to rule out implementations which do not deploy videos as the channel for out-of-class lecture transmission. As our definition follows closely to that of Bishop and Verleger (2013), the theoretical framework we adopt takes into consideration their perspectives. We shall elaborate on this theoretical framework in Section 2.

1.2 *Flipped classroom in Singapore schools*

In this past decade, there had been an emphasis for major shifts in classroom processes: (1) from lower-order thinking to higher-order thinking, (2) from analogue to digital, (3) from teacher-centred to student-centred learning, and (4) from isolation to collaboration on the part of the teachers' effort. As a result, an increasing number of teachers in Singapore schools have started experimenting flipped classroom to teach different subjects. Many of these attempts were carried out at small scale and at an exploratory stage. In comparison, a larger number of flipped classroom implementations took place in university settings in Singapore. Though there were emerging but isolated pockets of news articles reporting on the success of this approach when used in Singapore classrooms, there remains a paucity of formal research done in analysing the usefulness of flipped classroom in Singapore schools, and particularly in mathematics classrooms.

The authors speculate that the 'flipped classroom' educational approach may well continue for the next few years, so that developing an understanding of its relevance to mathematics teaching, in particular, seems to be a sensible research motivation. The principal purpose of this chapter is to fill a gap in the literature concerning the use of flipped classroom, together with its theoretical underpinnings, and its effectiveness in the context of Singapore, with particular emphasis in the content domain of mathematics. The authors set out to evaluate the effectiveness of an authentic implementation of flipped classroom in a Singapore junior college using testable propositions proposed by Abeysekera and Dawson (2015).

1.3 *Critical analysis through Self-Determination Theory (SDT) and Cognitive Load Theory (CLT)*

Abeysekera and Dawson (2015) recommended a critical analysis of the flipped classroom approach informed by two sound pedagogical theories: Self-Determination Theory (SDT) and Cognitive Load Theory (CLT).

SDT proposes that the basic cognitive needs for competence, knowledge and relatedness are universally applicable in learning. Students must feel competent to master the disciplinarity of the subject (e.g., content knowledge, skills, behaviours) that are required to be successful in a given social context (Ryan and Deci, 2000a). Based on SDT, it was suggested that flipped classroom might improve student motivation if it creates a sense of competence, autonomy and relatedness (Abeysekera & Dawson, 2015, p. 4). Such motivation can be of an intrinsic (i.e., those actions that individual engage in as they are inherently interesting and enjoyable) or an extrinsic nature (i.e., individuals engaging in actions because they lead to separable outcomes which are defined to be favourable outcomes distinct from inherent enjoyment).

CLT hinges on the notion of 'working memory' used by a learner when learning or problem solving (Miller 1956). It was reported therein that working memory consists of 7 ± 2 chunks. Thus, CLT informs classroom practitioners to be aware that because human working memory is subjected to certain kinds of load and thus overloading may in fact retard learning. Based on CLT, it was proposed that the flipped classroom approaches provide further facilities to distribute the cognitive load in such a way to make learning efficient (Abeysekera & Dawson, 2015, p. 8).

Through the lens of SDT and CLT, six propositions were put forth, (Abeysekera & Dawson, 2015) concerning the potential of flipped classroom in enhancing learner's motivation and managing learner's cognitive workload. These propositions are crafted in such a way that they are testable empirically by both quantitative and qualitative means. This paper can be regarded as an implementation to test these propositions. In Section 4, we elaborate on these propositions and explain how we design our study to test some of these propositions in our implementation.

2 Theoretical Framework for Flipped Classroom Pedagogy

Existing works on student-centred learning supported by learning theories that began with Piaget (1967) and Vygotsky (1978) already justified that classroom time must be spent in engaging students with meaningful learning activities. Because classroom time is at its premium, lectures and direct teaching when moved outside class can create this much needed time. Foot and Howe (1998) explained how constructivism and collaborative learning evolved from Piaget's theory of cognitive conflict, and co-operative learning stemmed from Vygotsky's zone of proximal development. It is noteworthy to point out Kolb's theory of experiential learning (Kolb & Kolb, 2012). Kolb established a theoretical model that consists of a universal learning cycle and two embedded dimensions: perception and processing – four learning styles are determined by all possible permutations of these two dimensions. Difference in learning styles therefore justify differentiated learning activities which are not possible to implement in traditional lecture-style setting.

However, we cautioned that merely moving from a traditional lecture to presenting the same lecture online is unlikely to result in better learning. To factor in the differences in learning pace and working memory (Clark, Nguyen & Sweller, 2005), we propose that students have flexibility in managing their cognitive workload. This can be done by pausing, rewinding, fast-forwarding or skipping parts of a video lecture to manage their own learning.

Bishop and Verleger (2013) modelled flipped classroom to be a system consisting of somewhat disjoint parts: interactive group learning activities inside the classroom, and direct computer-based individual instruction outside the classroom. Informed by SDT and CLT, our standpoint deviates from Bishop and Verleger (2013) in that flipped classroom pedagogy is far from being a disjoint sum of in-class human interaction and out-class computer-assisted learning. In order that flipped classroom be made effective, the teacher plays an important role in connecting these two aspects. More precisely, the teacher must purposefully design pre-class activities and post-class activities, frequently making changes that respond to the learning processes that take place either online or in-class. Figure 1 depicts the strong interplay between the interactive classroom activities and the explicit instruction methods assisted by media/computer technology which summarises the essence of flipped pedagogy.

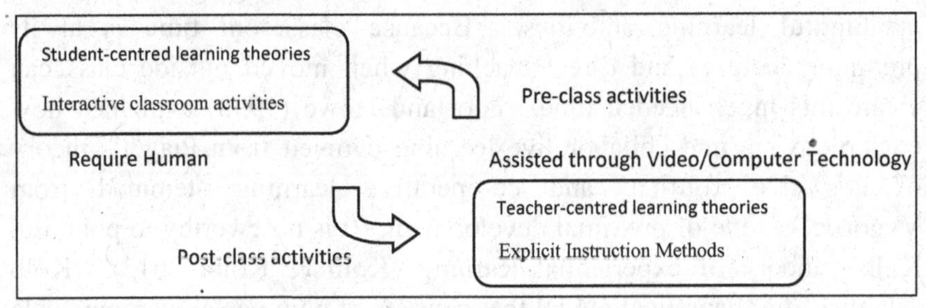

Figure 1. Theoretical framework for flipped classroom

3 Constructing the Flipped Classroom Package

The second author is a lecturer at a certain junior college (JC) in Singapore. JC students (JC1 are 11th graders; JC2 12th graders) who take mathematics as an 'A'-level subject are trained for two years in this subject and sit for a General Certificate Examination (Cambridge-Singapore syndicate) at the end of the final year. Mathematics is one of the core subjects required for entry to university. We classify JC education as tertiary education. Students from this JC were of mixed abilities and competencies in mathematics, with weaker students characterised by lack of interest, motivation and confidence in mathematics.

3.1 *Flipped classroom for Mathematics at junior college*

In the junior college, lessons are traditionally conducted via a lecture-tutorial system. JC2 students[1] would learn the contents of a new topic during the lecture, which is conducted thrice per week, 50-minutes per

[1] JC1 students had only two lectures per week, each lecture lasted 50 minutes.

lecture. Students would then duly complete tutorial exercises and clarify all their doubts with their tutors during the weekly tutorial lessons, comprising of three 50-minutes sessions.

During the lecture, it has been observed that many students who are stronger in the subject found the lecture pace rather slow. While the more motivated ones would do self-reading of the lecture notes and attempt the tutorial questions, the less-proactive fast learners would resort to sleeping or chatting during the lectures. On the other hand, students who are much weaker in the subject had problem coping with the pace of the lecturer. Not only did they fail to copy the necessary information to fill out the blanks in the lecture notes, they had problem grasping the new concepts taught, and thus faced much difficulty in attempting the tutorial questions.

There was thus a pressing need to explore an alternative pedagogy that would promote active learning, as well as to nurture self-motivated and independent learners so as to maximise learning during curriculum time.

3.2 *Description of the flipped classroom package*

Using a flipped-classroom approach and an adaptation of Team-Based Learning[2] (Haidet et al. 2012), three classes of students, totaling to 63[3], with varied learning abilities were taken out of the lectures to create time for group discussions. They were grouped into heterogeneous teams of 6 to 7 students in terms of their mathematics ability, gender and class.

The timetable of these classes was specially designed to accommodate two 100-minutes lessons per week (over 18 weeks, or

---

[2] We shall use the acronym TBL to stand for Team Based Learning. In a nutshell, the students in a class are formed into pre-assigned teams/groups according to the academic abilities and personalities. There was permanence in team-formation as the students remained in their assigned teams for two school terms (about 6 months).

[3] The flipped classroom started with 68 students in January but in the process 5 students downgraded the subject from H2 to H1 level, and thus dropped out of the flipped classroom.

equivalently two school terms) for mathematics, spanning from January to May. The sequence of learning activities is as follows:

(i) Pre-class reading: Students of these three classes were expected to self-learn the contents by reading the notes and filling out the blanks via any of the following options.
   (a) Referring to the PowerPoint Slides.
   (b) Referring to video-recording of the lectures.
   (c) Google for additional information to clarify their doubts.

(ii) In-class activities:
   (a) Individual Readiness Assurance Test (iRAT) was conducted at the beginning of the lessons when a new topic (or subtopic) is being introduced via online software which allows the teacher to obtain immediate feedback on the performance of all students. The area of weakness in content-understanding for each student, as well as the class as a whole, could be surfaced instantly. The students, however, would not get to know the results of the test immediately after taking the iRAT.
   (b) Group Readiness Assurance Test (gRAT) would follow immediately at the end of iRAT, using the same set of multiple-choice questions. Each team would discuss the questions, focusing on the justification of their choice of answer. Online software is used to allow each team to obtain immediate feedback on their selected answer. If the answer given was incorrect, the team had to re-deliberate their choice until a correct choice is made before moving on to the next question.
   (c) Class discussion – At the end of the gRAT, the teacher would facilitate the discussion of the MCQs to ensure that all students have gained basic understanding of the concepts learnt.
   (d) Applications of Concepts learnt – The class would proceed to solve challenging problem(s) by applying the concepts

learnt as a team and present the solutions or answers using mini-white boards or flash cards.

(e) Exit assessment – Conducted towards the end of a lesson to check each student's mastery of concepts and its applications. Feedback on the performance of the assessment would be addressed during the following lesson. Sometimes, due to time constraint, this exit assessment might be given as timed assignment to be submitted on the following day of the lesson.

(iii) Preparation for the next lesson – Students would attempt selected tutorial questions given in their lecture notes in preparation for discussion during subsequent lessons.

3.3 *Design considerations of flipped classroom package*

It is important to point out to the reader at this juncture that there were two groups of students: the control group (the rest of the classes other than the three chosen classes) underwent the traditional lecture-tutorial system, and the experimental group (the three chosen classes described in the preceding subsection) which underwent the flipped classroom system. The flipped classroom implementation must cover the same set of topics/subtopics as the traditional classroom implementation, which has been determined at the beginning of the year in the scheme of work as planned by the Mathematics Department of the participating JC. Pure mathematics topics include Vectors, Functions and Complex Numbers, while Statistics topics include Permutation and Combination, Probability, Binomial, Poisson and Normal Distributions, Sampling Theory, Hypothesis Testing, Correlation and Regression.

(i) PowerPoint slides – Every student of the cohort has a copy of the lecture notes with some of the solutions to the examples left blank. While students remaining in the lecture group would fill out the notes during lecture, students in the flipped-classroom experimental group were supported by specially designed

PowerPoint slides. The slides provided solutions to the examples with additional explanatory notes to aid understanding. In addition, there were pointers to highlight important concepts/common errors that require special attention of the students.

(ii) Video-recording of lectures – As the main lecture group was normally one or two lessons ahead of the flipped-classroom lesson pace in introducing a new topic (but their tutorial lessons were slower than the experimental group), the lectures were recorded using Camtasia and made available online. Students have the option of viewing the whole or part of the video to learn the contents of the topics. This allows the weaker students to have multiple viewings of the parts that they have problem with.

(iii) Readiness Assurance Test – For each introduction of a new topic or subtopic, students would begin the lesson with a readiness assurance test to ensure that they have done the pre-class learning. This is in the format of multiple-choice questions (MCQs). Each test comprises of about 5 to 10 questions that test the basic mastery of concepts and its simple application. At most one third of the MCQs may be set at the lowest cognitive level of Bloom's Taxonomy (Bloom, 1965), the knowledge recalling stage. The remaining questions would range from the comprehension, application to analysis stage of the taxonomy. The crafting of the MCQs need to be carefully planned to ensure its comprehensiveness in meeting all the learning objectives of the lesson. The distractors provided in each MCQ should be purposeful in surfacing a cognitive misconception/error in learning. The test consists of two parts, the Individual and the Group Readiness Assurance Test (iRAT & gRAT).

(a) iRAT – Conducted approximately at the pace of one minute per question.

(b) gRAT – using the same set of MCQs, its duration ranges from 10 to 20 minutes depending on the number of questions and the level of difficulties of the questions.

Figure 2 below shows an example of an iRAT/gRAT MCQ test item. Notice the non-routine element intended in the use of the *xz*-plane in the item.

The plane P is parallel to the xz-plane, and it passes through the point $A(1,2,3)$. The Cartesian equation of P is:

(A) $x + y + z = 6$ (B) $x = 1$ (C) $y = 2$ (D) $z = 3$

Figure 2. Sample MCQ item in iRAT/gRAT

(iv) Class discussion – To be facilitated by the teacher, exercising "no-hands-up" rule[4]. Pre-planned questions were posted to students who were pre-identified through the iRAT results. Good questioning techniques and facilitation skills are imperative here in guiding and prompting students through the discussion. The focus would be on getting students to

- Justify their choice of the correct answer
- Explain why the answer should not be his/her initial choice (The teacher needs to be tactful and not reveal the fact that the student being asked had given a wrong initial response. However, in reality, most students would "confess" that the wrong answer was his/her initial choice.)
- Share the group's points of discussion.

This practice allows the teacher to ensure that all students have gained basic understanding of the required concepts in the new topic.

---

[4] The "no-hands-up" policy is an Assessment-for-Learning questioning technique commonly used by teachers in Singapore classrooms, where the teacher asks a question to the whole class, and after giving sufficient wait time, the teacher selects any pupil to answer the question. The rationale behind this is that when children self-select themselves to contribute, the same set of children keep answering every time, depriving others a chance. (William, 2014)

(v) Application questions – This may include the tutorial questions of each chapter or other challenging questions to allow higher-order thinking in its application of concepts learnt. These questions are targeted at the higher cognitive level in Bloom's Taxonomy (Bloom, 1956), ranging from the application to evaluation stage. The questions must be challenging enough so that it would trigger intense team discussions. Each team would present their findings at the end of their discussion.

(vi) Exit Assessment – Consists of one or two questions similar to the application questions to assess students' learning of the lesson. Its duration is about 5 to 10 minutes. With prior notice of the exit assessment given at the start of the lesson, students generally tend to be more attentive in class and were engaged in the team discussions.

4 The Study

The aforementioned design of the flipped classroom package not only was aimed to address the motivational issues (lack of motivation to study mathematics and erosion of confidence as incompetency aggregated over time) experienced by students in this JC, as informed by SDT, but also to address issues concerning cognitive load of JC students (in general, a typical JC student manages with at least 3 subjects, but no more than 4, taken at the H2 level), as informed by CLT. We want to investigate the effectiveness of flipped classroom as an alternative teaching technique by testing some of the following propositions postulated by Abeysekera and Dawson (2015):

Proposition 1:
Learning environments created by the flipped classroom approach are likely to satisfy student needs for competence, autonomy and relatedness and thus, entice greater levels of intrinsic motivation.

Proposition 2/3/4:
Learning environments created by the flipped classroom approach are likely to satisfy students' need for autonomy/competence/relatedness and thus, entice greater levels of extrinsic motivation.

Proposition 5:
Student self-pacing of pre-recorded lectures may reduce cognitive load and help learning in a flipped classroom environment.

Proposition 6:
Flipped classroom approaches may provide more opportunities to tailor instruction to the expertise of students, enabling more appropriate management of the cognitive load.

4.1 *Intrinsic motivation versus extrinsic motivation*

According to SDT, intrinsic motivation is enhanced only when social contexts that *simultaneously* support feelings of competence, autonomy (Ryan & Deci, 2000a) and relatedness (Van Nuland, Taris, Bockaets, & Martens, 2012). In order for students to experience intrinsic motivation they must find engaging in a given learning activity inherently satisfying. According to Ryan and Deci (2000b), intrinsic motivation only takes place when learning activities are novel, challenging or supplies an aesthetic value for students. Notably, research informs us that

> "the freedom to be intrinsically motivated is found to decline as students move up from primary to tertiary education. Only a minority of students enrolled in contemporary higher education institutions are found to be intrinsically motivated. The vast majority are found to comprise students who are driven by extrinsic motivations due to increasing social demands from their personal and work lives." (Abeysekera & Dawson, 2015, p. 5)

For this reason[5], we assume that the majority of the JC students in this study are driven by extrinsic motivation. As a result, we have chosen not to address Proposition 1 in this chapter. However, we caution that extrinsic motivation does not equate to non-autonomy. The theoretical framework of SDT allows for differentiation in the degree of relative autonomy present in extrinsically motivated behaviour (Ryan & Deci, 2000a, 2000b). To illustrate this point, a student may complete the pre-class online activity because she understands that by doing so, she better prepares herself for the final year examination. Another student, wanting to conform to the school rules, also completes the same work. Both students are driven by extrinsic motivation; the first is of free choice while the second is in compliance with an external authority, and hence their difference in relative autonomy. Based on SDT, the flipped classroom approach can create learning environments that help students integrate certain associated values with the course as their own values via a process that is to be interpreted as a continuum. This means that on one end of the continuum, a student may begin with a state of unwillingness or merely going through motions (e.g., attend in-class activities because of compliance with school rules). At the other end of the continuum lies intrinsic motivation. So extrinsically motivated behaviours fall between these two ends and change according to the degree of autonomy of integration. Although intrinsic motivation is usually not experienced by tertiary students, SDT asserts that a form of extrinsic motivation that is characterised by a high relative autonomy, *integrated regulation*, is possible if the learning environment is designed to encourage it. When a student is in the state of integrated regulation, he has identified values associated with the given course and fully assimilated them to his self (Ryan & Deci, 2000a, 2000b).

---

[5] Apart from this justification, the assumption that we made here is supported by additional confidential information regarding the students of this participating JC, which cannot be disclosed herein.

4.2 *Autonomy, competence and relatedness*

There are three attributes associated to a learner's feeling when he or she engages in learning: autonomy, competence and relatedness. We briefly define each of these attributes as used in our ensuing discourse.

- Autonomy: The student *feels* in control and independent.
- Competence: The student *feels* competent to master the knowledge, skills and behaviours necessary to be successful in a given social context.
- Relatedness: The student gets a *sense* of belonging to a social group in a given context.

5 Method and Analysis

We designed a set of questions in the form of a summative survey to be participated by all students (a total of 63) who have agreed to be involved in this study. 56 students responded to the survey questions. Each question is crafted to measure the efficacy of the flipped classroom approach pertaining to the respective propositions (i.e., Propositions 2 to 6) in the mathematics classroom.

In Singapore, it is commonplace that mathematically weak students seek external help from private tutors. A majority 83% of those students who participated in the survey realised that they took ownership of their own learning *without* relying on private tutors. In contrast to past practice of not trying the tutorial questions before lesson, almost all (98%) students exercised their autonomy in choosing to attempt the designated tutorial questions before the TBL lessons. The statistics in Table 1 also indicates that these JC2 students (12th grade) were more likely to engage in learning, e.g., asked more questions as compared to the normal tutorial lessons. This willingness to engage was a phenomenal change because they were less engaged in the usual tutorial system and waited for the teachers to give them 'model answers'.

Table 1

Survey questions designed to test Proposition 2

Learning environments created by the flipped classroom approach are likely to satisfy students' need for autonomy and thus, entice greater levels of extrinsic motivation.

S/No.	Question	SA, A, D, SD*
1	Pre-class activities	
1Q5	I did the pre-class learning by (e) Having tuition (external help)	4%, 13%, 29%, 54%
1	Tutorial questions	
1Q7	(a) I attempted the designated tutorial questions before the TBL lessons.	41%, 57%, 2%, 0%
4	During the TBL lesson	
4Q2	(b) I was engaged in learning throughout the lesson	25%, 66%, 9%, 0%
4Q4	(d) I asked more questions as compared to normal tutorial lessons.	38%, 36%, 21%, 5%

*SA: Strongly agree, A: Agree, D: Disagree, SD: Strongly disagree

Of the participating students, 87% read the lecture notes to prepare for the TBL lessons as part of their pre-class activities (Table 2A, 1Q1). Because of this positive change, it was observed that the success rate of completing the designated questions was high (86% reported that they could do most of the questions). The students were most likely to be extrinsically motivated to read and prepare for the TBL lessons because they wanted to do well for the iRAT. The statistics in Table 2A also indicated that TBL discussions helped clarify their doubts and deepen their conceptual understanding. Notice that 87% of the students (Table 2A, 2Q3) did not just want the correct answer to the MCQs; rather their discussion led them to justify why an answer was correct. Also, the lecture notes were more thoroughly read by 85% of the students because of their discussion during TBL (Table 2B, 4Q8). Overall, these students gained confidence in mathematics.

Table 2A

Survey questions designed to test Proposition 3

Learning environments created by the flipped classroom approach are likely to satisfy students' need for competence and thus, entice greater levels of extrinsic motivation.

S/No.	Question	SA, A, D, SD
1	Pre-class activities	
1Q1	I did the pre-class learning by (a) Reading the lecture notes	55%, 32%, 9%, 4%
1	Tutorial questions	
1Q8	(b) I am able to do most of the designated tutorial questions before each lesson.	20%, 66%, 14%, 0%
2	Individual/Group Readiness Assurance Test	
2Q1	In general, the RAT (individual and group) help me gain a better understanding of basic concepts	59%, 38%, 3%, 0%
	During the gRAT,	
2Q3	(b) most of the time, we discussed the rationale for the choice of each option.	30%, 57%, 13%, 0%
2Q5	(d) I manage to clarify my doubts with my teammates.	43%, 55%, 2%, 0%
2Q6	(e) I gained a deeper understanding of the concepts involved through group discussion.	46%, 45%, 9%, 0%

Table 2B

Survey questions designed to test Proposition 3

Learning environments created by the flipped classroom approach are likely to satisfy students' need for competence and thus, entice greater levels of extrinsic motivation.

S/No.	Question	SA, A, D, SD
3	The discussion as a class	
3Q3	(c) further strengthened my own understanding of concepts.	46%, 50%, 4%, 0%
4	During the TBL lesson	
4Q3	(c) I was able to follow the lessons and understand the contents discussed.	27%, 57%, 16%, 0%
4Q7	(a) I gain a better understanding of the contents compared to lecture/tutorial lessons.	48%, 41%, 11%, 0%
4	On the whole, with TBL lesson	
4Q8	(b) I read my notes much more thoroughly than before.	46%, 39%, 13%, 2%

Having integrated the academic value of seeking to 'know the truth' in mathematics, 82% of the participating students were no longer satisfied with just the correct answer but they engaged in critical thinking while evaluating, defending or debating and defending viewpoints (Table 3, 2Q2). The sense of belonging to the community where mathematical discussions were encouraged motivated the participating students to participate in the discussion. In particular, a safe and respectful environment created for discussion during the TBL benefited the students as indicated by 93% who found that other groups contributed viewpoints which were missed out by their own group (Table 3, 3Q2). Because flipped classroom in this setting allowed the students to hear and be heard, the level of peer-to-peer relatedness these students experienced was increased.

Table 3

Survey questions designed to test Proposition 4

Learning environments created by the flipped classroom approach are likely to satisfy students' need for relatedness and thus, entice greater levels of extrinsic motivation.

S/No.	Question	SA, A, D, SD
2	Individual/Group Readiness Assurance Test	
2Q2	(a) most of the time, my group was only interested in what the correct answer is.	2%, 16%, 57%, 25%
2Q4	(c) we critically evaluate each other's viewpoint/suggestions.	38%, 48%, 14%, 0%
2Q7	(f) I had the opportunity to explain the concepts involved to my teammates.	43%, 55%, 2%, 0%
3	The discussion as a class	
3Q1	(a) helped to clarify doubts that my group had.	39%, 57%, 4%, 0%
3Q2	(b) revealed viewpoints of other groups that my group had omitted in our discussion.	39%, 54%, 5%, 2%
4	During the TBL lesson,	
4Q1	(a) My team worked well together	45%, 45%, 10%, 0%
4Q5	(e) I have the opportunity to explain my understanding of concepts to others.	27%, 50%, 21%, 2%
4Q6	(f) I get to listen to useful ideas or viewpoints of others which I have not thought of.	46%, 48%, 6%, 0%

About 39% of the participating students did not watch the LT1 lecture video. Instead, they read the PowerPoint slides (Table 4, 1Q3). The LT1 lecture videos were video recordings of the traditional lecture attended by the non-participating group. The participating students whose cognitive load for video watching could not exceed thirty minutes felt that the recorded video lectures were too slow-paced. Furthermore, the statistics in Table 4 indicates that the participating students were not motivated enough to access other resources (1Q4).

Table 4

Survey questions designed to test Proposition 5

Student self-pacing of pre-recorded lectures may reduce cognitive load and help learning in a flipped classroom environment.

S/No.	Question	SA, A, D, SD
1	Pre-class activities	
1Q3	(c) Watching the LT1 lecture video	18%, 43%, 32%, 7%
1Q4	(d) Referring to other resources e.g., relevant websites, textbooks	4%, 25%, 46%, 25%

It did not seem significantly more time-efficient for the participating students to handle the pre-class readings as compared to the traditional lecture-tutorial system. About 61% felt that they spent less time doing tutorial questions for TBL lessons compared to the lecture-tutorial system (Table 5, 1Q9). The design of the flipped classroom implemented here did not seem to provide significantly more opportunities to tailor instruction to the expertise of the students. In fact, watching videos could take up much more time for the out-class activities, and may add strain to cognitive load.

Table 5

Survey questions designed to test Proposition 6

Flipped classroom approaches may provide more opportunities to tailor instruction to the expertise of students, enabling more appropriate management of the cognitive load.

S/No.	Question	
1	Pre-class activities	
1Q6	The amount of time I spent in doing pre-class readings as compared to that spent by attending lectures and doing post-lecture readings are:	[Very much shorter/slightly shorter/ the same/slightly longer/very much longer] 27%, 18%, 13%, 21%, 21%
1	Tutorial questions	
1Q9	(c) The amount of time spent doing tutorials for TBL lessons is shorter than that in lecture/tutorial system.	SA, A, D, SD 21%, 40%, 25%, 14%

6 Findings

In this section, we report our findings from this present study of the efficacy of flipped classroom in teaching tertiary mathematics at the JC level. We do so by making evidence-based inference from three sources: (1) statistical analysis of the survey response (2) qualitative analysis of the free-response comments given by students in the survey, as well as (3) the comments given in selected students' interviews. Our analysis addresses each of the following aspects.

6.1 *Autonomy*

The participating students formed new habits of independently attempting the designated tutorial questions before the TBL lessons. In their process of answering these questions, the students either followed the recorded video lectures or the PowerPoint slides that explained the key notions via specifically designed examples. As the students viewed the electronically transmitted information, they actively filled in the blanks in the lecture notes.

Apart from wanting to score better in their iRAT, it was evident that the participating students felt that the flipped classroom approach had empowered them with the ability to learn mathematics independently:

> *Student 3: "TBL allow me to learn the chapter by myself [emphasis added] and I'm able to remember more things better since I've learnt a lot of things myself instead of just going to lecture and listening and just absorbing the things that I have learnt."*
>
> *Student 4: "moreover, given that now i learn the topic myself [emphasis added]..."*

Because of the perceived autonomy in learning, there was an overall increase of engagement in the lesson. The feeling of getting 'engaged' becomes a fuel for perseverance as the students progressed through the different topics.

> *Student 6: "... overall, I feel more engaged in the lesson and at the same time, ignited my passion in studying hard for the subject."*

Student 23 revealed in the interview that:

> *"Without the lecture, we have to study the topic on our own. ... Whenever I got my pre-class reading done, I learn a lot in the lesson."*

Participating students have departed from reliance on external help, for example employing private tutors or making prolonged consultation appointments with their mathematics teachers at school.

6.2 *Competence*

Competence can be displayed in various aspects of learning like mastering the studying skills, content-specific skills, problem-solving skills, communication skills, and so on. We observed that a majority of the students had studied their lecture notes more carefully, and made useful annotations on them, as exemplified in Student 24's sharing:

"To better prepare for the TBL lessons, we can watch the videoed lecture and go through the PowerPoint slides, and do annotations."

This practice increased the likelihood of solving the designated tutorial problems, a truly phenomenal change as contrasted with their past practice. We classified the motivation to gain genuine understanding as mastery of content knowledge. The Readiness Assurance Tests asserted positive pressure on the students because they created the needed platform for ensuing discussions, helping the students gain better understanding of the concepts. The structure of the pre-class activities provided scaffolding for the students, and aided in their resource management:

Student 1: "For me, I felt that it has helped to manage my time better [emphasis added] since a higher level of commitment is required in this form of learning."

As the students progressed towards integrated regulation, an increased sense of competence was also gained by sharing knowledge with students who did not participate in this method of learning:

Student 5: "... my friends from other classes often do not know how to answer. When I share it with them and it really make me feel that TBL is much better, rather than going to lectures and just listening."

Integrated regulation is characterised by a change in the student's belief system. In the past, the students did not believe in the rationale of practising mathematics when they studied at home. Now, the participating students are not only making sure everyone in the group had obtained the correct answers but the group strived to know how this answer was arrived at, as shown in Student 20's interview response:

"During the gRAT session, when we discussed the questions we ensured that all the members got their questions right. Not only did we ensure that all the members got the correct answer, but we also made sure we knew why the other three options were wrong."

From the students' responses, it was clear that as the students felt they have achieved competence in various skills, their motivation to study mathematics had progressed from the state of "merely going through motion" to that of genuine pursuit of knowledge, as shown in Student 20's interview response:

> *"For some people who prefer the lecture-tutorial system, it is because they just wanted to get the work done – which is not really something that I want. For me, learning is not just getting the work done, it is really about knowing your stuff."*

6.3 ***Relatedness***

The participating students critically evaluate each other's viewpoints and suggestions during their discussions. These discussions create opportunities for people to hear and to be heard, and hence satisfying this particular need of the individual.

> *Student 7: "I find my group very helpful in answering each other's doubts [emphasis added] and they are also very caring [emphasis added] by not allowing anyone to be left behind."*

An increased sense of relatedness among students and with the facilitating teacher emerged along time. As a result, there was a marked changed of practice in the classroom in that active discussions became a natural occurrence rather than an awkward enforcement on the part of the teacher. Meaningful group discussions also resulted in greater engagement and heightened attentiveness among the students, as Student 20 revealed:

> *"Not only did we clarify our doubts, but we also reinforced our own understanding by helping one another. When other raised doubts that I had not thought of, it instigates my curiosity which makes me more attentive in the session ... it is a very engaging method of learning as opposed to the conventional lecture-tutorial system."*

Another student, Student 23, pointed out TBL discussions were centred about purposefully crafted questions and scenarios, and these increased the students' attentiveness during the discussions and hence brought about greater engagement:

> *"Sometimes, in a [traditional] tutorial setting, we tend to zoom off because like a teacher is explaining the rest are listening, and sometimes the teacher will ask you questions and you answer. But this is not completely engaging. But then in a TBL lesson, the teacher gives you a scenario or maybe ask you a question to test your understanding, and you are forced to discuss with your friends. But when you are discussing with a friend, you are forced to stay attentive. That actually is beneficial."*

We observed from the survey that students grew in confidence and comfort, indicating that the teams had generally worked well.

> *Student 8: "TBL lessons really bring students from different classes together, cultivating group and teamwork [emphasis added]."*

However, we are also alerted about certain risks involved when the discussions could go the wrong direction:

> *Student 10: "Some of my teammates need to speak up more [emphasis added] and share whatever beneficial knowledge that can help us strengthen our foundation for topics."*
>
> *Student 2: "... a member in my TBL group has said condescending things to me before which really affected my learning and confidence [emphasis added]."*

Relatedness with the group is a double-edged sword that cuts both ways: either it encourages the student to engage deeper in meaningful discussions or repels the student from engaging because of other psychological effects. Students 24 and 20's comments shed some insight with regards to the preceding remark:

> *Student 24: "Basically, if you don't understand, you will look foolish pretending to explain something."*

> *Student 20: "Personally, I feel that TBL benefits me more as a person who needs to voice out my innermost doubts no matter how stupid they are. Because it is better to be a fool now than to be a fool in the exams, right?"*

6.4 *Cognitive load and its management*

From the survey response, there was no clear indication that the flipped classroom approach implemented in our study had taken sufficient consideration of the students' cognitive load involved in the learning, particularly in the pre-class activities. Indeed, there were several complaints about the efficiency (not the effectiveness) of viewing the videoed lectures and the PowerPoint slides prior to the TBL sessions.

> *Student 11: "... the ppt slides were really helpful in my understanding of concepts but the workload can be quite too much [emphasis added] at times."*
> *Student 9: "It's effective but the time (spent) is just too much [emphasis added]."*

Student 23 shared her difficulties in coping with a high cognitive load when it came to preparing her lessons:

> *"If I don't watch the lecture, I will have to take much longer to study a concept. So it was hard for me because I have external commitment for the first semester. Whenever I got my pre-class reading done, I learn a lot in the lesson. But when I did not manage to get things done, it was quite hard to stay on par with everybody."*

7 Implications and Conclusions

This chapter can be seen as responding to calling from Abeysekera and Dawson (2015) to test the efficacy of the flipped classroom approach via six testable propositions pertaining to motivational matters and cognitive load. By and large, our implementation showed positive effects in most aspects of extrinsic motivation afforded by the features put in place in our implementation.

While we are very encouraged by several phenomenal transformations concerning learner's motivation in mathematics in our implementation of flipped classroom approach that relied heavily on TBL, we must stay conscious of many potential risks, disadvantages or even dangers that are present.

In this study, we want to raise to the reader's attention that the design of the flipped classroom activities, both out-of-class or in-class, must be crafted with due consideration of the cognitive load of the students. Teachers who fore-load the lecture content as pre-class activities have a common tendency of making compact all that are needed to be covered in lectures and packing them as videoed lecture or reading materials. Students do need time and energy to 'unpack' these materials and to assimilate and/or accommodate them to their existing knowledge paradigm.

Perhaps, it is instructive to return to our theoretical framework for guidance in this matter. The role of the teacher is to continuously respond to the changes that are taking place in the class by designing *just-in-time* post-class activities (or pre-class activities prior to the next lesson) that address those changes or demands. Instead of anticipating too far ahead in the process of lesson planning, it is perhaps wiser to take the counter-intuitive approach of "wait-and-observe" so that a more responsive stance is possible.

Another important aspect that demands further study is the efficacy of group discussions. The effectiveness of group discussions in the TBL setting deserves a more careful study because the extent to which an individual finds group discussion beneficial depends on several social factors. Indeed some of these factors may well tie in with the culture or subculture of the group in question. Students who may have concerns of feeling humiliated may not find it comfortable to ask 'stupid questions' when there, in fact, is a need to ask such clarifying questions.

Acknowledgement

We would like to express our heartfelt appreciation to those JC2 students who agreed to participate in this study, and the participating JC, given that the JC2 students were sitting for the 'A' level examinations – a high-stake state examination.

References

Abeysekera, L., & Dawson, P. (2015). Motivation and cognitive load in flipped classroom: definition, rationale and a call for research. *Higher Education Research & Development, 34*(1), 1-14.

Berrett, D. (2012). How 'flipping' the classroom can improve the traditional lecture. *The Chronicle of Higher Education, 52*(8), 853-874.

Bishop, J. L. & Verleger, M. A. (2013). *The Flipped Classroom: A Survey of the Research.* The 120th American Society for Engineering Education Annual Conference & Exposition, June 23-26.

Bloom, B. (1956). Taxonomy of Educational Objectives, Handbook I: The Cognitive Domain. New York: David McKay Co Inc.

Clarke, R. C., Nguyen, F., & Sweller, J. (2005). *Efficiency in learning: Evidence-based guidelines to manage cognitive load.* San Francisco, CA: Pfeiffer.

Foot, H. & Howe, C. (1998). The psychoeducational basis of peer-assisted learning. In Topping K. J., & Ehly, S. W. (Ed.), *Peer-Assisted Learning,* 27-43. Lawrence Erlbaum Associates.

Haidet, P., Levine, R., Parmelle, D., Crow, S., Kennedy, F., Kelly, P. A., Perkowski, L., Michaelsen, L., & Richards, B. (2012). Perspective: Guidelines for Reporting Team-Based Learning Activities in the Medical and Health Sciences Education Literature. *Academic Medicine, 87*(3), p. 292-299.

Ho, W. K., Leong, Y. H, & Ho, F. H. (2015). The Impact of Online Video Suite on the Singapore Pre-service Teacher' Buying-in to Innovative Teaching of Factorisation via Algecards. In Ng, S. F. (Ed.), *Cases of Mathematics Professional Development in East Asia Countries – Using Video to Support Grounded Analysis*, 157-178. Singapore: Springer.

Kolb, A., & Kolb, D. (2012). Experiential Learning Theory. In Seel, N. M. (Ed.), *Encyclopedia of the Sciences of Learning*, 1215-1219. US: Springer.

Lage, M., Platt, G., & Treglia, M. (2000). Inverting the classroom: A gateway to creating an inclusive learning environment. *Journal of Economic Education, 3*(1), 30-43.

Miller, G. A. (1956). The magical number seven, plus or minus two: Some limits on our capacity for processing information. *Psychological Review, 63*(2), 81-97.

Piaget, J. (1967). *Six psychological studies.* New York: Random House.

Ryan, R., & Deci, E. (2000a). Intrinsic and extrinsic motivations: Classic definitions and new directions. *Contemporary Educational Psychology, 25*(1), 54-67.

Ryan, R., & Deci, E. (2000b). Self-determination theory and the facilitation of intrinsic motivation, social development, and well-being. *American Psychologist, 55*(1), 68-78.

Strayer, J. (2007). *The effects of the flipped classroom on the learning environment: A comparison of learning activity in a traditional classroom and a flip classroom that used an intelligent tutoring system.* Doctoral dissertation, The Ohio State University, Columbus.

Van Nuland, H., Taris, T., Bockaets, M., & Martens, R. (2012). Testing the hierarchical SDT model: The case of performance-oriented classrooms. *European Journal of Psychology of Education, 27*(4), 467-482.

Vygotsky, L. S. (1978). *Mind and society: The development of higher mental processes.* Cambridge, MA: Harvard University Press.

Chapter 13

Use of Comics and Storytelling in Teaching Mathematics

TOH Tin Lam   CHENG Lu Pien
JIANG Heng   LIM Kam Ming

The use of comics and storytelling in education has gained significant attention from researchers and educators worldwide over the past two decades. In this chapter, we describe an alternative package of teaching lower secondary mathematics topic on percentage and discuss how the use of comics in a storytelling setting, and with some help of technology, can be used in the mathematics classroom. This combination of comics and storytelling is an approach which can be used in teaching mathematics, especially to students who are less motivated and do not have high mathematical self-concept. The objectives and principles of designing the package as an alternative approach to the usual textbook material are discussed.

1 Introduction

It is generally agreed that modern education programs are more theory-based than skill-based (Glass, 2003). Thus it is not surprising to observe that modern education programs tend to favor the audio learners over the visual and kinesthetic learners (Amir & Subramaniam, 2007), who are usually associated with the less academically inclined students.

According to Myron and Keith (2007), teachers will be more successful in delivering classroom lessons if they are more mindful of the different learning styles and needs of their students and hence able to cater to the different learning needs of the individuals. We believe this is especially important if teachers are working with groups of students who are less motivated in the particular subject and who have low academic self-concept.

In an earlier survey conducted by Toh and Lui (2014), it was found that some Singapore teachers have already used cartoons, comics and storytelling to entice students who have low mathematics self-concept and who are less academically motivated to learn mathematics. However, there seems to be a lack of concerted effort among Singapore teachers to develop such material for school mathematics curriculum and researchers to study the impact of using this alternative approach in mathematics education in Singapore on students' mathematics self-concept and motivation to learn the subject.

This chapter reports the authors' work in developing an alternative package of teaching lower secondary mathematics topic of percentage using comics in the context of storytelling, the principles the authors adopted in designing the package and how this package could be executed in a secondary school mathematics classroom. Based on similar studies done overseas, the authors believe that this approach will likely have impact on students' motivation level and their mathematical self-concept. The research part is still work-in-progress and will be reported elsewhere in the future.

2 Use of Cartoons and Comics in Mathematics Classrooms

Cartoons are generally understood as two-dimensional visual art which are usually non-realistic or semi-realistic reflection of the real world, which are designed to create humor. Comics are media which convey ideas through visual images or series of images – these series of visual images are usually made up of cartoons. In this chapter, occasionally we use the two terms "cartoons" and "comics" interchangeably, bearing in mind that we are mainly interested in the use of visual images (which are usually humorous or exaggerated) to convey mathematical ideas.

School age children are generally attracted to cartoons and comics (Wright & Sherman, 2006). Students are generally comfortable in combining information in both visual and textual forms when reading comics. If comics are used in classroom teaching, they could logically provide opportunities for skill-building, creativity and reading for content (Urbani, 1978). The use of cartoons not only adds humor to learning, but also demonstrates the "human" side of the academic subject. Thus, the "enmity" between comics and school is gradually being dissolved as teachers are beginning to view comics as potential educational tools, as a way to motivate students in learning the various academic subjects (Cleaver, 2008). It is generally agreed that the use of comics can also be used to improve students' academic literacy (Tilley, 2008).

Undeniably, reading comics may not reach the level of complexity of reading texts in the real-world. However, "compared to reading 'real' books, reading comics appears to be a simple task and compared to reading no books, reading comics might be preferable" (Tilley, 2008).

With the increasing use of graphics in the society nowadays and the challenge the society is facing in representing large amounts of information in visual and graphic forms, it is crucial that students are exposed to alternative ways of representing and interpreting information using graphics instead of heavily worded statements (Lowrie, 2012). Samples of how comics can be used to bring across abstract mathematical concepts in a visual form, using cartoons as a powerful tool, are described in Toh (2009). In this way, the use of comics could be perceived as inducting students at a young age into the world of graphic representation in a casual and interesting way.

2.1 *What Research has shown about Cartoons and Comics*

There are several studies that described the impact of the use of cartoons and comics in teaching, both mathematics and other subjects. The use of cartoons and comics in the teaching of a topic in Chemistry at a secondary school had several positive outcomes: increased teacher-student and student-student interactions, increased students' interest, and higher levels of student participation in class (Lallbeeharry & Narod, 2014). In addition, the use of cartoons and comics as a teaching tool appeared to be useful in helping teachers to pinpoint students' problems and misconceptions about the topic, and in turn help improve students' understanding of the Chemistry topic and performance. Use of concept cartoons representing different viewpoints and correct and incorrect statements both facilitated students' engagement in argumentation and thinking. Lallbeeharry and Narod (2014) suggested that the use of cartoons is useful as a teaching tool that can be incorporated into a variety of classroom strategies such as group work and demonstration.

The study by Sengul and Dereli (2010) in Turkey on a group of 7th grade students shows that teaching mathematics with cartoons and comics lessened the students' mathematical anxiety. Similar studies by Sexton (2010) in Australia and Cho (2012) in the United States show further that 7th grade students preferred learning in the similar mode to the traditional mode of instruction. The use of cartoons and comics increased the students' motivation and interest in learning mathematics and decreased their mathematical anxiety.

Cartoons and comics are especially instrumental in arousing students' curiosity about the real world, especially if these can ride on the affordances of technology. There are two broad categories of the concept of curiosity: 1) sensory curiosity and 2) cognitive curiosity. Sensory curiosity which involves variation of sensory stimuli such as light, sound or other forms of stimuli may be best implemented through the use of multimedia technology (Liu, Toprac, & Yuen, 2009). Educators may also leverage on students' cognitive curiosity which is aroused when learners discover gaps in their knowledge base and they want to explore and learn with the use of technology (Malone & Lepper, 1987).

Cho, Osborne and Sanders (2015) found that the use of cartoons with a sample of elementary education pre-service teachers completing mathematics content courses led to positive outcomes such as increased enjoyment of the mathematics course and stronger engagement with the subject area. The use of cartoons also supported problem posing which has been linked to students' ability to think about and apply mathematics to real world situations.

3 Storytelling in Mathematics Classroom

Educators have been using storytelling to teach many academic subjects (Huber, Caine, Huber, & Steeves, 2013). Historically, mathematical concepts are discovered or invented when human beings considered solving real-world problems. However, it appears to many students that the mathematics in schools involves a communication of "unnatural" counterintuitive facts – which are usually represented by decontextualized symbolic mathematical language – and which is totally irrelevant to the real-world. Thus, it is essential to bridge the gap between the real-world mathematics and the school mathematics. Perhaps the use of storytelling in teaching mathematics could hold promise to bridge this gap.

Recently, mathematics teachers and educators have proposed ways to teach mathematics through storytelling (see, for example, Tan & Toh, 2013). In fact, some creative mathematics teachers, catering to the affective aspect of student learning, have used storytelling to make mathematics more enjoyable and, more importantly, to nurture a learning environment of imagination, emotion, and thinking. Stories are also used in mathematics classrooms not only to facilitate students to understand mathematical concepts, but also to engage them in activities, raise questions, think and explore various mathematical concepts and ideas (Schiro, 2004; Zazkis & Liljedahl, 2009).

Egan (2005) suggested that stories have their particular structure and must be able to appeal to our emotions:

> a story is a unit of some particular kind; it has a beginning that sets up a conflict or expectation, a middle that completes it, and an end that resolves it. The defining feature of stories, as distinct from other kinds of narratives – like arguments, histories, scientific reports – is that they orient our feelings about their contents. (p. 37)

According to Green (2004), "stories are a powerful structure for organizing and transmitting information, and for creating meaning in our lives and environments" (p. 1). Thus, the usefulness of using story in teaching engages students' emotions, spark their interest and help them create meaning connected to their lives.

One might worry that there is a possibility that students might get distracted in the process of engaging in stories during classroom lessons and thereby missed out the curriculum content. However, a careful design of the stories can make the lesson pedagogically sound in the mathematics classroom. The key principles in selecting appropriate stories in teaching are discussed in the next subsection.

3.1 *Principles in Selecting Appropriate Stories for Teaching*

Based on a survey of the mathematics education literature, we have identified two key principles in selecting and developing appropriate stories for teaching mathematics in the classroom.

Firstly, the selected stories must be able to "contextually situate(s) mathematics in ways that are interesting, involving, and relevant to the reader" (Schiro, 2004, p. 46). One key objective for teaching mathematics through storytelling is to engage the students to assume the role of the story's characters and "think through mathematical problems" (Schiro 2004, p.46). Thus, the key consideration here is that the students must be able to associate themselves with the context of the story so that we can achieve the objective to stir their emotions, intellect and imagination.

Secondly, the stories need to meaningfully bridge the mathematics concepts and the students' lives. Through the stories, meaningful mathematical tasks can be developed (as in the case of a word problem). These tasks should be "rich with mathematical possibility and opportunity, and contain hooks that connect the child's world with particular mathematical ideas and ways of thinking" (Ball, 1993, p. 375). For instance, the teacher Doris Lawson (in Schiro's study) used a fantasy story to help her students learn the multi-digit algorithm by having them "move" between the world of fantasy and the real world to solve the mathematical problems embedded in the story (Schiro, 2004).

4 Combining Comics and Storytelling

This section reports the authors' effort in combining the use of comics and storytelling to develop a mathematics teaching package on the lower secondary school mathematics topic of percentage. It describes how we develop the comic strips, the storytelling associated with the comic strips and the proposed way to deliver the lessons. The teaching package is made available to readers who wish to try out the teaching package at the website http://math.nie.edu.sg/tltoh/magical. The material is presented in the web-based form, in which practice questions are included at the appropriate juncture in the comic strips. Readers can click on the "practice" icons to access the practice questions. The teaching package can be used in both traditional print or web-based version, although the latter – with a little help from technology – has clearly more advantages than the former as discussed in the previous sections.

4.1 *Comics*

In developing the comics storytelling package in teaching the chapter on percentage in lower secondary school mathematics, the entire chapter is divided into five stories. Each of these five stories centres on the encounters of two characters Sam and Sarah in five episodes in their lives. Some incidents within these five episodes are realistic while others might be more exaggerated or even humorous. Instead of presenting the chapter on percentage with the conventional specific instructional objectives identified for each of the sub-units of the chapter, we classify the five episodes using titles of the comics (see Figure 1).

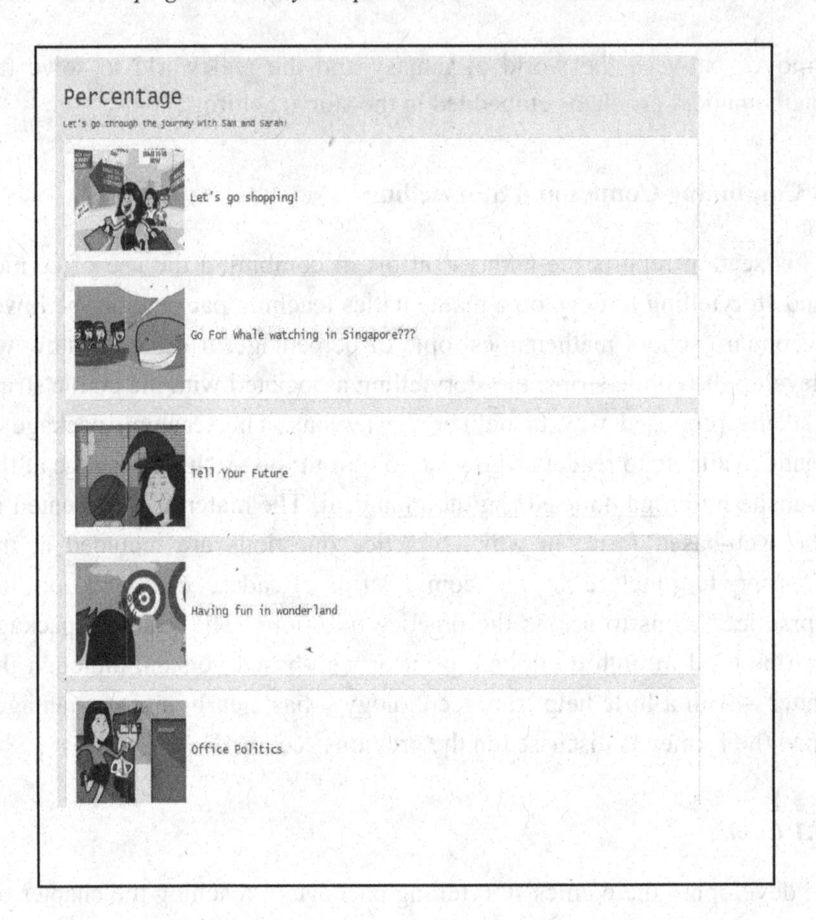

Figure 1. The titles of all the comics associated with percentage

Students are considered to have covered all the concepts in the chapter on percentage if they have read all the five sets of comics and attempted all the practice questions presented within them.

Each set of comics brings out the various key concepts within the chapter on percentage which students are required to be familiar with, as specified in the Singapore Mathematics Curriculum. We will next illustrate the first set of comics entitled "Let's go shopping", which consists of 18 windows and two practice sets. Each of the comic screens will be called a "Window" in our subsequent reference, as shown in Windows 1 to 18.

CRAZY SALE ON EVERY ITEMS!
Great Singapore Sale
70% SALE! COME TO US NOW
GREAT SALE 15% OFF STOREWIDE
30% OFF STOREWIDE!
1

COME TO US NOW
Waah! Many good buys and good news!
What does the symbol % mean?
30% OFF STOREWIDE!
SARAH
SAM
2

Veena Jeans
C Cut
That symbol % is the symbol for percent
Oh, what does it mean?
3
It means out of 100, so 30 % is 30 out of 100.
4

50 % SOLD!
5
What does this mean?
50 % SOLD!
6

50% means 50 out of 100. When reduced to the simplest form, $\frac{50}{100}$ is $\frac{1}{2}$. So 50% sold means $\frac{1}{2}$ of the merchandise sold.
7
Boat Ride
Another use of percentage.
SALON 50% DISCOUNT!
8

1
2
3
snip...
snip...
snip...
9
10

Why is your hairstyle so strange?
They only do 50% of hair cut!
11
Western Food ►
Boat Ride ►
hmmm.....
12

Yummy pizzas at 40% of the usual price!
13
14

sob...
sob...
15
I only pay 40% of the price!
Why did you only get $\frac{2}{5}$ of the pizza?
16

The first set "let's go shopping" attempts to bring out several concepts related to percentage through a shopping journey of Sam and Sarah, the main characters of the entire set of the comics. The following instructional objectives that are covered in the first set consist of:

- Interpret the use of the term "percentage" or the symbol %; and
- Express a percentage as a fraction.

These concepts or procedures are expressed through the conversation between the two characters Sam and Sarah, practice questions are embedded within suitable windows and opportunities to provide recall of pre-requisites are provided within the various sections of the comics. This will be elaborated in the next sub-section through our illustration with the first set "let's go shopping".

4.2 *Story associated with the Comic*

The set of comic strips is accompanied by a story describing the journey of Sam and Sarah. The story associated with "let's go shopping" is described below:

> Sam and Sarah are good friends. During the Great Singapore Sales they went down to the largest shopping centre in Singapore. They saw a young lady - a shopaholic - carrying many bags. (Windows 1 and 2)

> Sam was rather puzzled why the symbol % appears everywhere. Sarah said that % stands for percent. So she further said that 30% means 30 out of 100. (Windows 3 and 4)
>
> Sam and Sarah continued their shopping journey. See that? The shop indicated that 50% sold! (Windows 5 and 6)
>
> So 50% means 50 out of 100. Can it be expressed as a fraction? Yes it should be $\frac{50}{100}$. Sarah explained that $\frac{50}{100}$ can be expressed as $\frac{1}{2}$. (Windows 7 and 8)
>
> Sam and Sarah continued their shopping and came across a hair salon. The salon offered 50% discount! They saw a young boy coming out of the salon and only got half of his hair cut. Do you know why? (Windows 9 to 12)
>
> They were hungry after a long day of shopping. Outside a pizza shop, Sam and Sarah saw that pizzas were sold at 40% of the usual price. This sounds like a good deal. He thought that he could have the whole pizza by paying 40% of the price! But see, what did he get? He was so disappointed that he only got $\frac{2}{5}$ of the pizza. Do you know why? (Windows 13 to 18).

The story begins with a visual image to get students to appreciate the use of percentage in the real world. Visual symbols of the percentage sign (%) are used in the context of a shopping centre. A brief discussion serving as a recap of percentage as "out of 100" is next introduced (Windows 4 to 8). A humorous way of introducing the conversion of an "easy percentage" of 50% as $\frac{1}{2}$ was introduced through the haircut incident (Windows 9 to 12). A more in-depth thought provoking incident of engaging students to think of how 40% is equivalent to $\frac{2}{5}$ is presented next (Windows 13 to 18). At Window 18, the teacher could facilitate students to discuss how percentages can be converted to fractions (in their simplest forms), demonstrate with examples and explanations and get their students to practice the appropriate questions provided within the package.

4.3 *Support for teaching*

The students have been introduced to the elementary concept of percentage at the upper primary level. The chapter on percentage (on which the comics and storytelling material was designed and described in this chapter) taught at the lower secondary level serves to build on the students' fundamental concepts learnt at the primary level. Thus, the support provided within the teaching package includes opportunities for students to recall the meaning of percentage (Figure 2) and practise some elementary questions on percentage (Figure 3) before they move on to the comic strips.

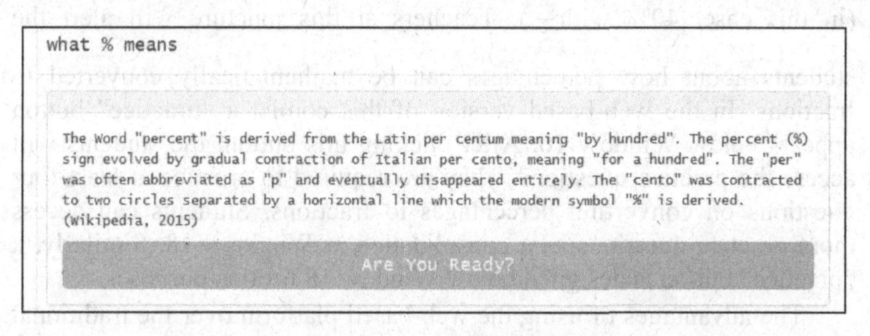

Figure 2. A recap station on the meaning of percentage (%)

At Window 2, students are given the opportunity to recall the meaning of percentage. A "recall" button is designed below Window 2. Upon clicking this button, Figure 2 appears. This screen serves to facilitate students to recall the meaning of percentage. In addition, students can appreciate the historical meaning of the word "percentage" and how the symbol (%) was derived. Students are given further opportunity to answer some elementary questions about percentage if they are ready to proceed, as indicated by the question "Are You Ready?"

An example of the practice questions corresponding to key learning objectives of the chapter on percentage is shown in Figure 3.

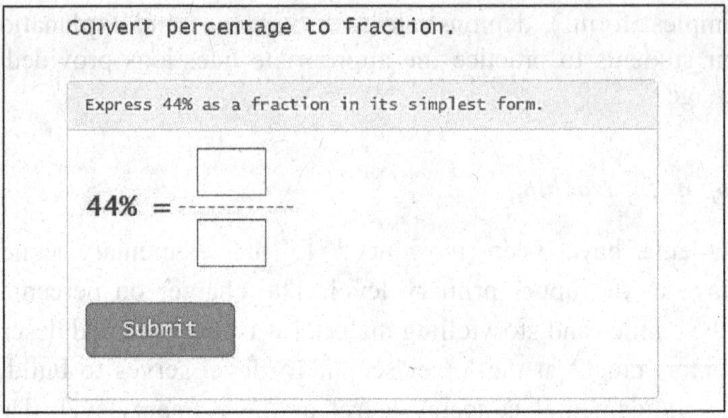

Convert percentage to fraction.

Express 44% as a fraction in its simplest form.

44% = ----------

Submit

Figure 3. Practice questions associated with converting percentage to fraction

Window 16 presents the idea of associating a percentage with a fraction (in this case, 40% with $\frac{2}{5}$). Teachers at this juncture will alert the students about how percentages can be mathematically converted to fractions. In the web-based version of this comic, a "practice" button appears below Window 16. After clicking this button, the students will access the practice questions. They are required to practice at least four questions on converting percentages to fractions. Students can access more practice questions as a consolidation at Windows 18. Similarly, a "practice" button is designed below Window 18 for this purpose.

The advantages of using the web-based platform over the traditional printed version are that (1) the numbers in the questions can be randomized so that the students have the opportunity to practice different questions each time they access the website; and (2) the system can be programmed such that students are given immediate feedback, especially in the event that the answers provided by the students are incorrect.

4.4 *Trialling with the comics storytelling package*

The authors first developed one story that described the entire chapter of percentage. Comic strips were developed to tell the story with practice questions introduced at the appropriate junctures of the story. This inaugural set of comics was shared during several seminars with school mathematics teachers for their suggestions. The suggestions given by the teachers are:

- It is not advisable to use one long story to cover the entire topic on percentage, as there is a high chance that students may get bored with the same context for too many lessons. It is desirable to chunk the entire story into several shorter sets of comic strips with different story settings in order not for the students to feel the novelty of the comics so as not to get bored.
- It is encouraged to include some humour into the comic strips to sustain the students' interest. The humour should target to introduce mathematical concepts or ideas in a casual manner.
- If technology is to be used in this comic storytelling approach, the material should be interactive; students should be given immediate feedback in their responses to answering questions in the practice sections.

In our revision of the first draft of comic storytelling package, we incorporated the above points. The comic strips, whose content is shown in Figure 1, is a fine-tuning of our first draft based on the authors working on the teachers' suggestions.

4.5 *Lesson implementation*

There are several possible approaches to conduct a lesson using comics and storytelling as the pedagogy. We shall describe two possible approaches.

Firstly, the teacher can choose to go through the story as described in Section 4.2, interjecting questions to elicit students' responses to the story. As teachers ask students questions, the students will activate their *schema* and thereby they will have the opportunity to assimilate the new learning into their existing structure. Furthermore, in telling a story, it is not the mathematical precision that matters; what is even more crucial is to impress the related mathematical concepts deeply into the mind of the students.

Secondly, student role play can be used; instead of the teachers reading the stories verbatim to the students, these stories can be acted, performed vividly and discussed in the mathematics classroom which honours various voices and perspectives (Whitin & Whitin, 2000). Students can get more interested in the content of discussion if they can associate themselves with the characters of the story and, further, in thinking about their own strategies and feel more comfortable sharing their ideas to solve the mathematical problems.

5 Conclusion

This package was originally developed as part of a research project in helping low attainers learn mathematics. Students will get interested in the learning processes if they are able to make connections between the real-world and the related mathematical concepts (Albert and Antos, 2000). The use of cartoons and other visuals also demonstrate to the students the "human" part of the subject besides its relevance to the real world.

We believe that appropriate use of cartoons and comics in storytelling, is able to motivate students in mathematics, especially the less academically inclined. The use of cartoons and visual representations will appeal to the visual learners, while the use of technology to allow the students to navigate through the entire set of comics will reach out to the kinesthetic students. Although our target groups of students are those who are not motivated to learn the subject or those with low academic self-concept, we believe that the same pedagogy can motivate students to learn mathematics among the general student population.

From a pragmatic perspective, there will likely be more emphasis on using graphic stimulus in the curriculum and school practices in aligning to the increasing use of graphics in the society (Lowrie, 2012). An additional advantage of using comics in teaching (in addition to capturing the students' interest about the topic) is that it nurtures the development of the skill of interpreting real-world information presented in graphic form. The use of comics storytelling pedagogy certainly enriches the learning experience of students, although there is currently little research in this area of combining the effects of comics and storytelling.

Using comics and storytelling in teaching mathematics facilitates and builds students' confidence in learning mathematics. Not only that, it allows students to be creative in their outlook in learning mathematics and enjoy learning mathematics. Cartoons and comics have a tendency to exaggerate the real-world situation while at the same time it is enjoyable for the students in the process of learning. This provides a good opportunity for students to develop critical thinking skills and to appreciate diverse views. Consequently, we are developing the students to be self-directed learners. We are envisioning that the students will eventually take ownership of their own learning, and even design their own comics to convene mathematical ideas – in other words, they become active contributors to the learning processes. It is clear that these are critical components of the 21st century competencies.

Acknowledgement

The authors would like to thank our consultant Dr Elena Lui Hah Wah for her invaluable suggestions on this paper, and all school teachers who have given precious suggestions and comments in the various phase of the development of the teaching package. This development-cum-research work is supported by a development grant from the National Institute of Education (DEV 7/14 TTL), of which the authors are eternally grateful.

References

Albert, L., & Antos, J. (2000). Daily journals. *Mathematics Teaching in the Middle School, 5* (8), 526–531.

Amir, N., & Subramaniam, R. (2007). Making a fun Cartesian diver: A simple project to engage kinaesthetic learners. *Physics Education, 42*(5), 478-480.

Ball, D. (1993). With an eye on the mathematical horizon: Dilemmas of teaching elementary school mathematics. *The Elementary School Journal, 93*, 373-397.

Cho, H. (2012). *The Use Cartoons as Teaching a Tool in Middle School Mathematics.* ProQuest, UMI Dissertations Publishing.

Cho, H., Osborne, C., & Sanders, T. (2015). Classroom experience about cartooning as assessment in pre-service Mathematics content course. *Journal of Mathematics Education at Teachers College, 6*(1), 45-53.

Cleaver, S. (2008). Comics & graphic novels. *Instructor, 117*(6), 28-30.

Egan, K. (2005). *An imaginative approach to teaching.* San Francisco: John Wiley & Sons, Inc.

Glass, S. (2003). *The uses and applications of learning technologies in the modern classroom: Finding a common ground between kinaesthetic and theoretical delivery.* Educational Research Report. Information Analyses (070).

Green, M. C. (2004). Storytelling in teaching. *Observer*, *17*(4), 1-7.

Huber, J., Caine, V., Huber, M., & Steeves, P. (2013). Narrative inquiry as pedagogy in education: The extraordinary potential of living, telling, retelling, and relieving stories of experience. *Review of Research in Education, 37*(1), 212-242.

Lallbeeharry, H., & Narod, F. B. (2014). An Investigation into the Use of Concept Cartoons in the Teaching of "Metals and the Reactivity Series" at the Secondary Level (pp. 41-66). In M. G. Bhowon, S. Jumeer-Laulloo, H. L. K. Wah, & P. Ramasami (Eds.), *Chemistry: The key to our Sustainable Future.* New York: Springer.

Liu, M., Toprac, P., & Yuen, T. (2009). What factors make a multimedia learning environment engaging: A case study. In R. Zheng (Ed.), *Cognitive effects of multimedia learning* (pp. 173-192). Hershey: PA: Idea Group.

Lowrie, T. (2012). Visual and spatial reasoning: The changing form of mathematics representation and communication. In B. Kaur, & T. L. Toh (Eds.), *Reasoning, communication and connections in mathematics* (pp. 149 – 168). Singapore: World Scientific.

Malone, T. W., & Lepper, M. (1987). Making learning fun: A taxonomy of intrinsic motivations for learning. In R. E. Snow, & M. J. Farr (Eds.), *Aptitude, learning, and instruction: III. Conative and affective process analyses* (pp. 223-253).

Myron, H., & Keith, H. (2007). Advice about the use of learning styles: A major myth in education. *Journal of College Reading and Learning, 37*(2), 101-109.

Schiro, M. (2004). *Oral Storytelling and Teaching Mathematics: Pedagogical and Multicultural Perspectives.* Thousand Oaks: CA: SAGE.

Sengul, S., &. Dereli, S. (2010). Does instruction of "Integers" subject with cartoons effect students' mathematics anxiety? *Procedia – Social and Behavioral Sciences, 2*, 2176–2180.

Sexton, M. (2010). Using concept cartoons to access student beliefs about preferred approaches to mathematics learning and teaching. *Annual Meeting of the Mathematics Education Research Group of Australasia.* Western Australia: Mathematics Education Research Group of Australasia.

Tan, S. H., & Toh, T. L. (2013). On the teaching of the representation of complex numbers in an argand diagram. *Learning Science and Mathematics Online Journal, 8*(1), 75-86.

Tilley, C. L. (2008). Reading comics. *School Library Media Actiivities Monthly, 24*(9), 23-26.

Toh, T. L. (2009). Use of cartoons and comics to teach algebra in mathematics classrooms. In D. Martin, T. Fitzapatrick, R. Hunting, L. C. Itter, T. Mills, & L. Milne (Eds.), *Mathematics Of Prime Importance: MAV Yearbook 2009* (pp. 230 - 239). Melbourne: The Mathematical Association of Victoria.

Toh, T. L., & Lui H. W. E. (2014). Helping normal technical students with learning mathematics - A preliminary survey. *Learning Science and Mathematics Online Journal, 10*(1), 1-10.

Urbani, T. (1978). Fun, funny, funnies. *Teacher, 96*(1), 60-68.

Whitin, P., & Whitin, D. J. (2000). *Math is language too: Talking and writing in the mathematics classroom.* Urbana, IL: National Council of Teachers of English, and Reston, VA: National Council of Teachers of Mathematics.

Wright, G., & Sherman, B. (2006). Comics redux. *Reading Improvement, 43*(4), 165-172.

Zazkis, R., & Liljedahl, P. (2009). *Teaching Mathematics as Story-telling.* SENSE Publishers.

Chapter 14

Game Theory: An Alternative Mathematical Experience

Ein-Ya GURA

Few branches of mathematics have been more influential in the social sciences than game theory. In recent years, it has become an essential tool for social scientists studying the strategic behavior of competing individuals, firms, and countries. However, the mathematical complexity of game theory is often very intimidating for students who have only a basic understanding of mathematics. We address this problem here by offering a course in game theory (specifically one topic) that can be adapted to the classroom to help students understand the key concepts and ideas of game theory without the use of formal mathematical notation.

1 Introduction

In the twenty-first century, game theory stands in the forefront of interdisciplinary research and yet this branch of mathematics is completely ignored in high-school mathematics curricula. Game theory undertakes to build mathematical models and draw conclusions from these models in connection with interactive decision-making situations in which a group of people not necessarily sharing the same interests are required to make a decision.

Mathematical modeling is the essence of the teaching of mathematics and it is recognized as an important process skill in the mathematics curriculum (MOE, 2012). The problem is that much of the mathematics taught at school has no connection to real-life situations and therefore the mathematical modeling is not very meaningful. Game theory is a branch of mathematics that is motivated mostly by the social sciences or, better, by human behavior and therefore constructing mathematical models for real-life situations is natural to the theory. In game theory one can ensure that students are able to understand the basic problem on which they are working. As game theorist and Nobel laureate Robert Aumann writes

> The language of game theory – coalitions, payoffs, markets, votes – suggests that it is not a branch of mathematics, that it is motivated by and related to the world around us, and that it should be able to tell us something about the world. (Aumann, 1985, p. 28)

Yet, despite its tractability, game theory is a branch of mathematics: "the resistive medium is the mathematical models with its definitions, axioms, theorems and proofs" (Aumann, 1985, p. 42). As it was said, game theory is largely absent in the mathematics classroom. To address this issue, a course in game theory was designed. Our main goal in designing this course was to make mathematics a subject that can be discussed and thought about through a basic comprehension of the problem at hand. We wanted to present material that does not require mathematical prerequisites and yet involves deep game-theoretic ideas and some mathematical sophistication.

2 The Course

The course is a collection of a few topics from game theory that are intended to open a window onto the new and fascinating world of mathematical applications to the social sciences. It takes a small number of topics and studies them in depth. It shows the student how a mathematical model can be constructed for real-life issues. One of the aims of the course is to acquaint the student with "a different mathematics," a mathematics that is not buried under complicated formulas, yet contains deep mathematical thinking. Another aim is to show that mathematics can efficiently handle social issues. A third aim is to deepen the mathematical thinking of the person. Details about the various topics can be found in the book by Gura and Maschler (2008).

In the next section, we will describe one example from the course which can be adapted to the classroom, namely Mathematical Matching. The topic of Mathematical Matching is exemplified by the problem of assigning applicants to institutions of higher learning. Each applicant ranks the universities in which he is applying according to his scale of preferences. The universities, in turn, rank the applicants for admission according to their own scale of preferences. The question is how to effect the "matching" between the applicants and the universities. The problem leads to unexpected solutions.

3 Mathematical Matching

In 1962 a paper whose title, "College Admissions and the Stability of Marriage," raised eyebrows (Gale & Shapley, 1962). Actually, the paper dealt with a matter of some urgency. According to Gale (2001) the paper owes its origin to an article in the New Yorker, dated September 10, 1960, in which the writer describes the difficulties of undergraduate admissions at Yale University. Then as now, students would apply to several universities and admissions officers had no way of telling which applicants were serious about enrolling. The students, who had every reason to manipulate, would create the impression that each university was their top choice, while the universities would enroll too many students, assuming that many of them would not attend. The whole process became a guessing game. Above all, there was a feeling that actual enrollments were far from optimal. Having read the article, Gale and Shapley collaborated. First, they defined a concept of stable matching, and then proved that the stable matching between students and universities always exists.

For simplicity, Gale and Shapley started with the unrealistic case in which there are exactly n universities and n applicants and each university has exactly one vacancy. A more realistic description of this case is a matching between men and women—hence the title of their paper.

The Matching Problem: Consider a community of men and women where the number of men equals the number of women.

Objective: Propose a good matching system for the community (the meaning of "good" will become clear presently). To be able to propose such a system, we shall need relevant data about the community. Accordingly, we shall ask every community member to rank members of the opposite sex in accordance with his or her preferences for a marriage partner. We shall assume that no men or women in the community are indifferent to a choice between two or more members of the opposite sex. This assumption is introduced to simplify our task.

We illustrate with Example 1: that of four men named Al, Bob, Cal and Dan, and four women Ann, Beth, Cher and Dot. Their list of preferences is given in Figure 1. The numbers indicate what rank a man or a woman occupies in the order of preferences. The left numbers indicate the men's preferences and the right numbers indicate the women's preferences. For example, according to the men's ranking of women, Al ranks Cher first, Dot second, Ann third, and Beth last. And according to the women's ranking of men, Cher ranks Bob first, Cal second, Al third and Dan last.

	Ann	Beth	Cher	Dot
Al	3,1	4,1	1,3	2,2
Bob	2,2	3,2	4,1	1,3
Cal	1,3	2,3	3,2	4,1
Dan	3,4	4,4	2,4	1,4

Figure 1. List of preferences in Example 1

Now given each individual's preferences, can we propose a matching system for the entire community?

Proposal 1:

(Al – Dot, Bob – Ann, Cal – Beth, Dan – Cher)

2 x 2   2 x 2   2 x 3   2 x 4

The numbers below each couple indicate what rank one member of a couple assigns to the other member. The number on the left indicates what rank the man assigns to the woman; the number on the right, what rank the woman assigns to the man. This is indeed a possible proposal, but it is not a good one. Cher is displeased, because she is paired off with her last choice. She can propose to Bob, but she will be turned down because she is his last choice. She will fare no better with Cal, because she is his third choice while he is paired off with his second choice. On the other hand, if Cher proposes to Al, he will be very pleased, because she is his first choice. Proposal 1 is not a good one because Cher and Al prefer each other to their actual mates, and one can reasonably assume that they will reject the matchmaker's proposal.

Let us consider another proposal where we try to pair each man with his first choice. Al chooses Cher and Cal chooses Ann. However, both Bob and Dan would pick Dot. Between the two of them, Dot would prefer Bob. This leaves Dan to be paired with Beth. We summarize the situation below.

Proposal 2:

(Al – Cher, Bob – Dot, Cal – Ann, Dan – Beth)

1 x 3   1 x 3   1 x 3   4 x 4

Now three out of four men are paired off with their first choice. Would Proposal 2 be accepted or rejected?

Conversely, we can ask if it is possible to pair off all the women with their first choice. Specifically, both Ann and Beth prefer Al, while Cher's first choice is Bob and Dot's first choice is Cal. We see that if we pair off Ann with her first choice, Al (since Al prefers Ann to Beth), then Beth cannot be paired with him too. We cannot pair off Beth with her second choice, Bob, as he is already paired off with Cher. And Beth's third choice, Cal, is already paired off with Dot. Beth is therefore left with her last choice.

Proposal 3:

(Al – Ann, Dan – Beth, Bob – Cher, Cal – Dot)

3 x 1 4 x 4 4 x 1 4 x 1

Three of the four women are paired off with their first choice. Will they accept or reject this matching? Perhaps not since Beth can contest this matching. For example, she can approach Bob and suggest that they both reject this matching and form their own pair. In so doing Beth gets her second choice—better than her fourth choice—and Bob gets his third choice—better than his fourth choice. Thus, the above matching will be rejected by both Beth and Bob.

Both Proposal 1 and Proposal 3 were rejected for the same reason and can be regarded as unstable. We can thus propose that a stable matching system must satisfy the following requirement: *A stable matching system must be such that under it there cannot be found a man and a woman who are not paired off with each other but prefer each other to their actual mates.*

Based on the above definition, we can see that Proposal 2 is stable. As Al, Bob and Cal are already paired with their first choice, they would not protest. The only man who can do better is Dan, but every woman other than Beth prefers their current mate to Dan.

It is natural to ask if it is always possible to find a stable matching in any preference system. The answer is yes and Gale and Shapley (1962) proposed a procedure for finding a stable matching system. The Gale–Shapley algorithm is based on three assumptions:

- The number of men equals the number of women.
- There is no indifference.
- Every community member has to rank all members of the opposite sex.

We shall illustrate the algorithm with Example 1. The algorithm involves a number of stages. In the first stage, each man proposes to the woman who is his first choice. Each woman with more than one proposal chooses her favorite and rejects the rest. The chosen man is considered to be on her waiting list. Specifically, Al proposes to Cher, Bob proposes to Dot, Cal proposes to Ann and Dan also proposes to Dot. Since Dot has two proposals, she chooses Bob and rejects Dan. Thus Al, Bob and Cal are respectively on Cher, Dot and Ann's waiting lists. This concludes the first stage. In the next stage, each man who is not on any waiting list proposes to the most preferred woman whom he had not yet proposed to. Each woman gets to choose between the new proposals and the one on waiting list. The chosen one is placed on her waiting list, while all others are rejected. The process is then repeated until every man is on some woman's waiting list. Returning to our example, Dan now proposes to Cher. Between Dan and her waiting list candidate, Al, Cher would prefer Al and reject Dan. This ends the second stage. Dan now proposes to his third choice Ann in stage 3. Again, Ann prefers Cal (currently on her waiting list) and rejects Dan. Finally in stage 4, Dan proposes to Beth who places him on her waiting list. We now have a complete matching which is actually Proposal 2.

The Gale–Shapley algorithm guarantees that a stable matching can always be found. It is proved by the theorem: *The Gale-Shapely algorithm terminates in a stable matching system.* It has been shown that some preference structures yield more than one stable matching system. This raises a few questions. For example, *is there one stable matching system that is everyone's favorite?* Assuming that there is no indifference, the answer is no, because if there are two stable matching systems, then at least one man is paired off with a different woman in the second system, and necessarily prefers one system to the other. Another question is the following. *Is there one stable matching system that is the men's favorite?* Surprisingly, the answer is yes. The same goes for the women: there is a stable matching system that is the women's favorite. This leads to a definition of an optimal stable matching system for all men or all women. It can be proved that for every preference structure, the matching system obtained by the Gale–Shapley algorithm, when the men propose, is optimal for the men. When the women propose they too get an optimal matching system. We illustrate with Example 2 (Figure 2). It is clear that when the men propose, the algorithm results in the stable matching of Eric with Esther and Fritz with Fannie. But when the women do the proposing, the stable matching would be Eric with Fannie and Fritz with Esther.

	Esther	Fannie
Eric	1,2	2,1
Fritz	2,1	1,2

Figure 2. List of preferences in Example 2

Let us now apply the Gale-Shapley algorithm to Example 1 again but this time with the women proposing. In Stage 1, we have both Ann and Beth choosing Al, Cher choosing Bob and Dot choosing Cal. Since Al prefers Ann to Beth, Ann is placed on his waiting list while Beth is rejected. We represent this stage with the following diagram. The asterisk (*) indicates Beth is rejected while the other three women are on waiting lists.

Stage 1:

Al	Bob	Cal	Dan
Ann	Cher	Dot	
Beth*			

In Stage 2, Beth, the only one not on a waiting list, now proposes to Bob who is second on her preference list. Between Beth and Cher, Bob chooses Beth and rejects Cher.

Stage 2:	Al	Bob	Cal	Dan
	Ann	Cher*	Dot	
		Beth		

In Stage 3, Cher is now the only one not on a waiting list. She proposes to Cal, who chooses her over Dot.

Stage 3:	Al	Bob	Cal	Dan
	Ann	Beth	Dot*	
			Cher	

For Stages 4 and beyond, the choices are represented by the following diagrams.

Stage 4:	Al	Bob	Cal	Dan
	Ann*	Beth	Cher	
	Dot			

Stage 5:	Al	Bob	Cal	Dan
	Dot	Beth*	Cher	
		Ann		

Stage 6:	Al	Bob	Cal	Dan
	Dot	Ann	Cher*	
			Beth	

Stage 7:	Al	Bob	Cal	Dan
	Dot*	Ann	Beth	
	Cher			

Stage 8:	Al	Bob	Cal	Dan
	Cher	Ann*	Beth	
		Dot		

Stage 9:	Al	Bob	Cal	Dan
	Cher	Dot	Beth*	
			Ann	

Finally in Stage 10, since Beth has been rejected by Al, Bob and Cal, she proposes to her fourth choice Dan who accepts her on his waiting list. The algorithm terminates here and we arrive at the stable matching identical to Proposal 2.

Stage 10:	Al	Bob	Cal	Dan
	Cher	Dot	Ann	Beth

We see that both the men propose and the women propose version of the algorithm leads to the same stable matching in Example 1. It can be proved that this implies the stable matching is unique (Gura & Maschler, 2008).

Gale and Shapley were the first to ask whether their algorithm for matching men and women was applicable to the college admissions problem. What they did not know at the time was that the Association of American Medical Colleges had already for ten years been applying the Gale–Shapley algorithm to the task of assigning interns to hospitals in the United States (Roth & Sotomayor, 1990). By a process of trial and error that spanned over half a century, the Association in 1951 adopted the procedure that was hospital-optimal. We also note that Shapley and Roth subsequently shared the 2012 Nobel Prize in Economics science for contributions to the theory of stable matchings.

4 Conclusion

In this chapter, we described an example in game theory that can be adapted to the classroom. We started with a real-life problem, and made some simplifying assumptions to render the problem tenable. Several possible solutions are discussed, which lead to the definition of a stable matching. An algorithm for obtaining a stable matching is then presented. It is hoped that such a discussion would pique the interests of students in game theory. Those wishing to learn more, including the proofs of the various claims, can refer to the book by Gura and Maschler (2008). In the book, the authors went on to show how the three assumptions of the Gale-Shapley algorithm could be dispensed with and how the algorithm could be adapted to deal with the original admission problems. The other three chapters in the book consider other real-life situations where constructing a mathematical model can be relevant and meaningful.

References

Aumann, R. J. (1985). What is game theory trying to accomplish? In K. Arrow, & S. Honkapohja (Eds.), *Frontiers of Economics*. Oxford: Basil Blackwell.

Gale, D., & Shapley, L. S. (1962). College admissions and the stability of marriage. *American Mathematical Monthly*, *69*, 9-15.

Gale, D. (2001). The two-sided matching problem: origin, development and current issues. *International Game Theory Review*, 3, 237-252.

Gura, E-.Y., & Maschler, M. (2008). *Insights into Game Theory*. Cambridge: Cambridge University Press.

Ministry of Education, Singapore. (2012). *O & N(A)-Level Mathematics Teaching and Learning Syllabus*. Singapore: Author.

Roth, A.E., & Sotomayor, M. (1990). *A Study in Game-Theoretic Modeling and Analysis*. Cambridge: Cambridge University Press.

Contributing Authors

Divya BHARDWAJ is a research associate at the Mathematics and Mathematics Education Academic Group of the National Institute of Education (NIE), Singapore. Since 2011, she has been involved with various research projects at the Centre for Research in Pedagogy and Practice and the Mathematics and Mathematics Education Academic group at NIE. Prior to that, she was a secondary school mathematics teacher at NPS International School, Singapore. She completed her Master of Science (Mathematics for Educators) from NIE in 2009 and the Cambridge International Diploma for Teachers & Trainers in 2010. Currently, she is pursuing a Doctor of Philosophy in Mathematics.

CHAN Puay San graduated from the National University of Singapore with a Bachelor of Science (Hons) in Mathematics. In 2009, she was conferred the degree of Master of Education (Curriculum and Teaching) by the Nanyang Technological University, Singapore, under the Ministry of Education Postgraduate Scholarship. A dedicated educator who believes in bringing out the best in every child, as well as a reflective practitioner who constantly seeks improvement in teaching pedagogy, she received the President's Award for Teachers in 2013. In recognition of her outstanding contribution and devoted service, she was awarded the National Day Award Commendation Medal (Pingat Kepujian) in 2015.

CHENG Lu Pien, received her PhD in Mathematics Education from the University of Georgia (U.S.) in 2006. She is a lecturer with the Mathematics and Mathematics Education Academic Group at the National Institute of Education (NIE), Nanyang Technological University, Singapore. She specialises in mathematics education courses for primary school teachers. Her research interests include the professional development of primary school mathematics teachers, tools and processes in mathematics education programmes for pre-service teachers. Her research interests also include children's thinking in the mathematics classrooms.

CHUA Boon Liang is an Assistant Professor in mathematics education at the National Institute of Education, Nanyang Technological University in Singapore. He holds a PhD in Mathematics Education from the Institute of Education, University College London, UK. His research interests cover pattern generalisation, mathematical reasoning and justification, and task design. Given his experience as a classroom teacher, head of department and teacher educator, he seeks to help mathematics teachers create a supportive learning environment that promotes understanding and inspire their students to appreciate the beauty and power of mathematics. With his belief that students' attitudes towards mathematics are shaped by their learning experiences, he hopes to share his passion of teaching mathematics with the teachers so that they make not only their teaching more interesting but also learning mathematics an exciting and enjoyable process for their students. He feels honoured to have been awarded Excellence in Teaching by the National Institute of Education in 2009 and 2013.

DINDYAL Jaguthsing is an Associate Professor in the Mathematics & Mathematics Education Academic Group at the National Institute of Education, Nanyang Technological University in Singapore. He teaches mathematics education courses to both pre-service and in-service teachers. He has worked on teacher education projects and currently has specific interest in teachers' use of examples in the teaching of mathematics. His other interests include the teaching and learning of geometry and algebra, lesson study and students' reasoning in mathematics specifically related to their errors and misconceptions.

Ein-Ya GURA is a member of The Federman Center for the Study of Rationality at The Hebrew University of Jerusalem. The Hebrew University's Center for Rationality and Interactive Decision Theory is a unique venture in which faculty, students and guests join forces to explore the rational basis of decision-making. The range of the Center's activities is unparalleled in the world. Whereas most interdisciplinary centers aim to promote cooperation between researchers in two or three different fields, the Center for Rationality is a truly multidisciplinary enterprise, drawing on the talents of outstanding scholars from 13 different departments, in four faculties of the University. Among the Center's members are top scholars from the fields of mathematics, economics, psychology, biology, political science, education, computer science, philosophy, business, statistics and law. At the core of the Center's work is Interactive Decision, or Game Theory. Game theory models are virtually limitless in the ways they can be applied to the real world. As part of the Center's activities Ein-Ya Gura conducts a course in game theory for high-school students credited by the university. Once a week the students come to the Center for a 3 hours meeting to study game theory and to listen to enrichments lectures given by members of the Center. Before retirement Ein-Ya Gura was the head of the mathematics department at the Rothberg International School at the Hebrew University and at the same time she was a senior lecturer of mathematics.

Keiko HINO is Professor of Mathematics Education at Utsunomiya University in Japan. She received her M.Ed. from Tsukuba University and Ph.D. in Education from Southern Illinois University. She began her career as a Research Assistant at Tsukuba University in 1995. After the career as Associate Professor of Mathematics Education at Nara University of Education, she is now working at Utsunomiya University. Dr. Hino has been Professor since 2010 at Utsunomiya University. Her major scholarly interests are students' development of proportional reasoning and functional thinking through classroom teaching, international comparative study on teaching and learning mathematics, and mathematics teachers' professional development. She has published authored or co-authored 2 books, 20 book chapters, and over 40 journal articles and presented at over 40 conferences, including the International Congress on Mathematics Education, the International Conference of Psychology of Mathematics Education, the East Asia Regional

Conference on Mathematics Education, and annual meetings of the Japan Society of Mathematics Education and of Japan Society for Science Education. She is also involved in activities for improving mathematics education as an editor of Japanese Primary and Lower Secondary School Mathematics Textbooks and External-expert for Lesson Study in Mathematics.

HO Weng Kin received his Ph.D. in Computer Science from The University of Birmingham (UK) in 2006. His doctoral thesis proposed an operational domain theory for sequential functional programming languages. He specializes in programming language semantics and is dedicated to the study of hybrid semantics and their applications in computing. Notably, he solved the open problem that questions the existence of a purely operationally-based proof for the well-known minimal invariance theorem of (nested) recursive types in Fixed Point Calculus. His research interests also include domain theory, exact real arithmetic, category theory, algebra, real analysis and applications of topology in computation theory, as well as teaching of pre-university and tertiary mathematics via technologically-based pedagogies.

JIANG Heng is an assistant professor in the National Institute of Education, Nanyang Technological University, Singapore. She obtained her PhD in Curriculum and Instruction from Michigan State University (U.S.) in 2011. Her expertise includes teacher learning; qualitative research in education; teacher performance assessment; and curriculum studies. She has published in peer-reviewed academic journals such Journal of Teacher Education, Intercultural Education, and Frontier of Education in China; and delivered presentations in international conferences. Her recent book is “Learning to Teach with Assessment: A Student Teaching Experience in China” by Springer in 2015.

Berinderjeet KAUR is a Professor of Mathematics Education at the National Institute of Education in Singapore. Her primary research interests are in the area of classroom pedagogy of mathematics teachers and comparative studies in mathematics education. She has been involved in numerous international studies of Mathematics Education and was the Mathematics Consultant to TIMSS 2011. She is also a member of the MEG (Mathematics Expert Group) for PISA 2015. As the President of the Association of Mathematics Educators (AME) from 2004-2010, she has also been actively involved in the professional development of mathematics teachers in Singapore and is the founding chairperson of Mathematics Teachers' Conferences that started in 2005 and the founding editor of the AME Yearbook series that started in 2009. She was awarded the Public Administration Medal by the President of Singapore in 2006.

Barry KISSANE is an Emeritus Associate Professor in Mathematics Education at Murdoch University in Perth, Western Australia. From 1985 until his recent retirement, he taught primary and secondary mathematics teacher education students at Murdoch University, except for a period for which he was Dean of the School of Education and an earlier period working and studying at the University of Chicago. His research interests in mathematics education include the use of technology for teaching and learning mathematics and statistics, numeracy, curriculum development, popular mathematics and teacher education. He has written several books and many papers related to the use of calculators in school mathematics, and published papers on other topics, including the use of the Internet and the development of numeracy. Barry has served terms as President of the Mathematical Association of Western Australia (MAWA) and as President of the Australian Association of Mathematics Teachers (AAMT). He has been a member of editorial panels of various Australian journals for mathematics teachers for around 30 years, including several years as Editor of The Australian Mathematics Teacher. A regular contributor to conferences for mathematics teachers throughout Australasia and elsewhere, he is an Honorary Life member of both the AAMT and the MAWA.

KOAY Phong Lee is a senior lecturer at the National Institute of Education in Singapore. Currently she is teaching mathematics education courses to both pre-service and in-service teachers. She has written textbooks and developed teaching resources for primary school teachers.

Her research interests include learning difficulties in mathematics, mathematical investigations and problem-solving.

NG Lay Keow is a Senior Teacher in Mathematics, Ministry of Education, Singapore. She holds a master's degree in Mathematics Education, from the National Institute of Education, Nanyang Technological University in Singapore and is currently studying for her doctoral degree. She has been teaching Mathematics in a Secondary School for eight years and her work entails mentoring beginning teachers, building professional capacity and driving research within the school.

LIM Kam Ming is the Associate Dean for Programme Planning and Management (Office of Teacher Education) at the National Institute of Education (NIE), Nanyang Technological University in Singapore. He is also an Associate Professor with the Psychological Studies Academic Group at NIE. His research interests include prosocial behaviour, help-seeking behaviour, locus of control, self-concept and motivation. He was conferred the Public Administration Medal by the President of the Republic of Singapore in 2015. He received the Award for Outstanding Contribution to Psychology in Singapore from the Singapore Psychological Society in 2005. He is currently the President of the Educational Research Association of Singapore (2015–2017).

Cynthia SETO is a Principal Master Teacher (Mathematics) at the Academy of Singapore Teachers, Ministry of Education. She holds a PhD in mathematics education and her dissertation is on classroom learning environment. She leads the Mathematics Chapter to deepen teachers' pedagogical content knowledge and mathematical knowledge for teaching. Her research interests include metacognition, mathematical modelling, mentoring and teacher learning communities. She has published and presented papers at national and international conferences. With more than thirty years of classroom teaching experience, she has taught all levels of mathematics from Secondary Two to Primary One. She has received several innovation awards, such as Microsoft-MOE Professional Development Award, Hewlett-Packard Innovation in

Information Technology Award, for teaching of mathematics using technology.

Stephen THORNTON is Executive Director of Mathematics by Inquiry, an Australian government funded project conducted by the Australian Academy of Science in collaboration with the Australian Association of Mathematics Teachers. The project produces classroom and professional resources to promote inquiry approaches to school mathematics from Foundation to Year 10. Steve has previously lectured in mathematics teacher education at the University of Oxford, the University of Canberra and Charles Darwin University. He has wide experience as a teacher of mathematics, as a researcher in mathematics education, as leader of a national professional development program for the Australian Mathematics Trust, and as a consultant and critical friend to numerous mathematics education projects. Steve is a Life Member of the Australian Association of Mathematics Teachers and has been awarded a B.H.Neumann medal for excellence in Australian mathematics education. He has written numerous journal articles and book chapters, and has presented keynote addresses and workshops at Australian and international mathematics education conferences. Steve holds B.Sc. (Hons) and Grad. Dip. T. degrees from the University of Adelaide; and a Ph. D. from the Australian National University.

TOH Pee Choon received his PhD from the National University of Singapore in 2007. He is currently an Assistant Professor at the National Institute of Education, Nanyang Technological University. A number theorist by training, he continues to research in both Mathematics and Mathematics Education. His research interests include problem solving, the teaching of mathematics at the undergraduate level, as well as the use of technology in teaching.

TOH Tin Lam is currently the Deputy Head of the Mathematics and Mathematics Education Academic Group in the National Institute of Education, Nanyang Technological University, Singapore. He obtained his PhD from the National University of Singapore in 2001. Dr Toh continues to do research in mathematics as well as mathematics education. He has papers published in international scientific journals in both areas.

WONG Lai Fong has been a mathematics teacher for over 20 years and is known for her efforts in engaging students with fresh and creative strategies in the study of Mathematics. For her exemplary teaching and conduct she was given the President's Award for Teachers in 2009. She sets the tone for teaching the subject in Anderson Secondary School, being Head of Department (Mathematics) from 2001 to 2009 and currently a Lead Teacher for Mathematics. She is also an executive committee member of the Association of Mathematics Educators, and a member of the Singapore Academy of Teacher's Math Chapter Core Team. She was awarded a Post-graduate Scholarship by the Singapore Ministry of Education to pursue a Master of Education in Mathematics which she has completed in 2014. Currently, she is involved in several Networked Learning Communities looking at ways to infuse mathematical reasoning, metacognitive strategies, and real-life context in the teaching of mathematics. Lai Fong is also active in the professional development of mathematics teachers and in recognition of her significant contribution toward the professional development of Singapore teachers, she was awarded the Associate of the Academy of Singapore Teachers in 2015.

YEO Kai Kow Joseph is a Senior Lecturer in the Mathematics and Mathematics Education Academic Group at the National Institute of Education, Nanyang Technological University, Singapore. As a teacher educator, he is involved in training pre-service and in-service mathematics teachers at primary and secondary levels and has also conducted numerous professional development courses for teachers in Singapore and overseas. Before joining the National Institute of Education in 2000, he held the post of Vice Principal and Head of Mathematics Department in secondary schools and also served a stint in the Ministry of Education, Singapore. His research interests include mathematical problem solving in the primary and secondary levels, mathematics pedagogical content knowledge of teachers, mathematics teaching in primary schools and mathematics anxiety.

WONG Khoon Yoong has enjoyed working as a mathematics educator at four universities in four countries over the past four decades. During this period, he taught courses in mathematics education, provided consultancy to institutes in several countries, published widely, and participated in the review of the national mathematics curriculum of three countries. His latest publication was entitled *Effective Mathematics Lessons through an Eclectic Singapore Approach*, the 2015 yearbook of the Association of Mathematics Educators. Since his retirement from the National Institute of Education in July 2014, he continues to work as a part-time consultant for Singapore schools and institutes. He may be contacted at khoon.y.wong@gmail.com.

Printed in the United States
By Bookmasters